Collins

Exploring Physics

GRADE 9

Derek McMonagle

Series editor: Marlene Grey-Tomlinson

Collins

HarperCollins*Publishers* Ltd
The News Building
1 London Bridge Street
London SE1 9GF

First edition 2017

10 9 8 7 6 5 4 3 2 1

ISBN 978-0-00-824419-4

www.collins.co.uk/caribbeanschools

A catalogue record for this book is available from the British Library.

Typeset by QBS Learning
Printed in Italy by Grafica Veneta SpA

Author: Derek McMonagle
Series editor: Marlene Grey-Tomlinson
Illustrators: QBS Learning
Publisher: Elaine Higgleton
Commissioning Editor: Tom Hardy
Project Manager: Alissa McWhinnie
Editor: Alexander Rutherford
Copy Editor: David Hemsley
Proofreader: Niamh O'Carroll and Mitch Fitton
Cover image: Lana Po/Shutterstock

MIX
Paper from responsible sources
FSC™ C007454
www.fsc.org

This book is produced from independently certified FSC™ paper to ensure responsible forest management.

For more information visit: www.harpercollins.co.uk/green

Derek McMonagle is a leading writer of science educational materials with worldwide experience. He has developed courses at primary, secondary and advanced levels for many countries including Jamaica and the UK.

Marlene Grey-Tomlinson is a Senior Teacher at Excelsior High School in Kingston, Jamaica. She has been teaching Integrated Science and Biology in Grades 7 though to 11 for over 25 years and is one of the country's leading science educationalists.

Contents

Introduction – How to use this book 2

Unit 1: Working like a physicist 4

Unit 2: Static electricity 14

Unit 3: Current electricity 22

Science in practice – lighting system for a model house 48

Unit 4: Magnetism 50

Science in practice – building an ammeter 70

Unit 5: Thermal energy 72

Science in practice – using fuel made from recycled material instead of charcoal 100

Science in practice – making a cool box 102

Index 104

Acknowledgements 107

Introduction – How to use this book

Grade 9

3.2 Electricity and safety

We are learning how to:
- distinguish between electrical insulators and conductors
- use electricity safely.

This tells you what you will be learning about in this lesson.

Electricity is potentially dangerous so we must learn how to use it safely.

Some plugs have two pins and some have three. Sockets may have caps to prevent children poking things into them.

FIG 3.2.1 An electrical appliance is connected to the mains electricity supply by a plug, which fits into a socket

The book has plenty of good illustrations to put the science into context.

Activity 3.2.1

Examining a plug and socket

Here is what you need:
- plug
- socket.

SAFETY

Even though plugs and sockets are made of plastic, they should never be touched with wet hands. There is a danger that wetness will pass into the plug or socket and this can result in an electric shock.

Here is what you should do:

1. Look carefully at the plug and observe where the wires from the appliance will be connected.
2. Notice that when the plug is placed fully in the socket there are no exposed meta parts.
3. Look carefully at the socket and where the wires from the supply will be connecte
4. Notice that the holes for the pins of the plug are very small, which prevents peopl accidentally pushing something into the holes.
5. Notice also that the plug pins must be fully into the socket before they connect w the electricity supply. This means that no parts of the plug pins are exposed.
6. From what type of material are the bodies of the plug and socket made?

24 Exploring Physics: Unit 3 Current electricity

3.2

Wires in an electrical circuit

Two wires are needed to make a circuit. Some plugs have three pins and three wires. The third wire is called the **earth wire**. It protects the user should a fault develop in the appliance.

If the wires in a metal table lamp came loose and touched the metal body, anyone touching it would receive an electric shock. The earth wire is connected to the metal body. Most of the current will pass through it so the user would get a small shock and not come to any harm.

FIG 3.2.2 Metal-bodied appliances should be earthed

An electrical appliance that has a plastic body does not need to be earthed because plastic is an insulator. Such appliances are described as **double insulated**.

FIG 3.2.3 Plastic-bodied appliances do not need to be earthed

There are often some interesting facts.

Check your understanding

1.

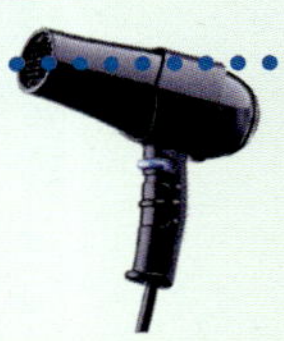

FIG 3.2.4 Gabriella's hairdryer

a) How does Gabriella connect her hairdryer to the power supply?
b) Why should Gabriella not use her hairdryer when her hands are wet?
c) What is the outer casing made of?
d) Will Gabriella's hairdryer have a 2-pin or a 3-pin plug? Explain your answer.

Each spread offers questions to help you to check whether you have understood the topic.

Interesting fact

Electricians use a detector to locate electrical wiring in walls and floors to avoid accidentally drilling through a cable and getting an electric shock.

Key terms

earth wire wire in a plug that is connected to the metal body of the appliance to protect the user in the event of a fault

double insulated an appliance that has a body made of an insulating material

Key terms are defined on the pages where they are used.

25

Review of Current electricity

- Electricity is a convenient form of energy.
- Materials that conduct an electric current are called electrical conductors.
- Materials that do not conduct an electric current are called electrical insulators.
- Metals are good conductors because they contain delocalised electrons. Their particles carry electrical charge along the metal. Insulating materials do not contain delocalised electrons.
- Electricity is potentially dangerous so electrical devices must be handled with care.
- Electrical appliances with metal bodies are connected to an earth wire. This provides a pathway for current in the event of wires coming loose and touching the metal body, which protects the user from a severe electric shock. Electrical appliances with plastic bodies do not require an earth wire and are said to be double insulated.
- A cell is a means of providing a small and safe amount of energy. A battery is formed when two or more cells are joined together.
- A cell has a positive (+) terminal and a negative (–) terminal. When cells are placed together to form a battery, they must all be pointing in the same direction.
- In order to flow, an electric current must have a complete circuit.
- An electric current can be measured using an ammeter. The unit of current is the ampere, A, or for smaller currents the milliampere, mA. There are 1000 milliamperes in 1 ampere.
- A circuit diagram is a method of representing a circuit by a series of connected symbols. Each symbol represents a component in the circuit. Circuit diagrams are quick to draw and will be universally understood because scientists around the world use the same symbols.
- Components in a circuit might be connected in series or in parallel.
- When lamps are connected in series, the more lamps there are the dimmer they are and the less current flows in the circuit.
- When lamps are connected in parallel, they shine with the brightness of a single lamp up to the point where the cell or battery cannot provide any more electrical energy.
- Lamps connected in parallel are brighter than lamps connected in series but they draw more current from the battery. This means that the battery will be exhausted more quickly. Lamps connected in series will shine less brightly but will continue to shine for longer because the battery will last longer.
- Resistance is measured in ohms and has the symbol Ω.
- Resistors can be used to control the flow of current in a circuit.
- There is a difference in potential energy or voltage between the terminals of a cell and this is expressed in volts.
- Voltage is measured in volts using a voltmeter.
- When the voltage across a circuit changes the current in it will also change.

44 Exploring Physics: Unit 3 Current electricity

At the end of each group of Units there are pages which list the key topics covered in the Units. These will be useful for revision.

Science, Technology, Education and Mathematics (STEM) activities are included which present real-life problems to be investigated and resolved using your science and technology skills. These pages are called **Science in practice.**

At the end of each section there are special questions to help you and your teacher review your knowledge. See if you can apply this knowledge and if you can use the science skills that you have developed.

Review questions on Current electricity

1. Redraw the circuit in Fig 3.R.1 as a circuit diagram, using appropriate symbols for the components.

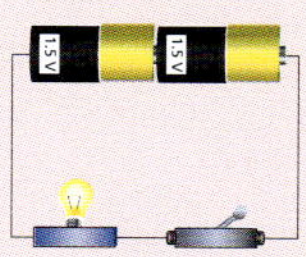

FIG 3.R.1

2. Dante used the circuit shown in Fig 3.R.2 to test whether materials conduct electricity.

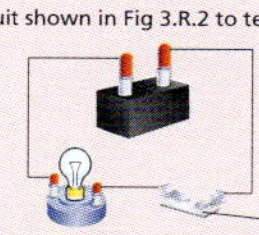

material FIG 3.R.2

a) How was Dante able to decide whether or not each material conducted electricity?
b) Before testing the materials, Dante connected the two crocodile clips together. Why did he do this?
c) Here is a list of the materials Dante tested. Arrange them in groups according to whether or not they conduct electricity.

copper cardboard plastic iron
wood glass steel lead

d) What name is given to a material that does not conduct electricity?
e) Give one common feature of all of the materials that conduct electricity.

45

Science in practice

Lighting system for a model house

Before you start work on this activity, you should recall that mains electricity is dangerous. You should not remove the tops from lamps and switches or expose wires in mains lighting circuits.

Mr Livingston is making a model doll house as a surprise for his daughter's birthday.

FIG SIP 3.1 The model doll house Mr Livingston is making

He wants to make it extra special by having a light in the ceiling of each room and a switch by the door that can turn it on and off, but he doesn't know anything about electricity.

1. You are going to work in a group of three or four to design a lighting system for Mr Livingston's doll house, to make the components and to make a model to show how your system works. The tasks are:
   - to research about house lighting
   - to design a suitable lighting circuit
   - to design and make lamps and switches to a suitable scale for a doll house
   - to build a model of a doll house out of cardboard
   - to install your lighting system in your model
   - to test your lighting system
   - to consider how your lighting system might be modified or extended.

a) Take a look at how the lighting is arranged in your home.

In each room the switch only controls the light for that room.

Use your knowledge of series and parallel circuits to design a suitable lighting circuit for a doll house that contains four rooms.

FIG SIP 3.2 Room light and switch

48 Exploring Physics: Unit 3

Unit 1: Working like a physicist

1.1 Area of regular shapes

We are learning how to:

- formulate a simple working definition for the term 'area'
- use appropriate units for area
- calculate the area of regular shapes

A grid is a set of squares, like a sheet of graph paper. In Fig 1.1.1 each of the small squares measures 1 cm by 1 cm.

The **area** of a flat shape is the space contained within its boundaries. The area contained by a square of sides of 1 cm is called a square centimetre or a **centimetre squared**. It has the symbol **cm²**.

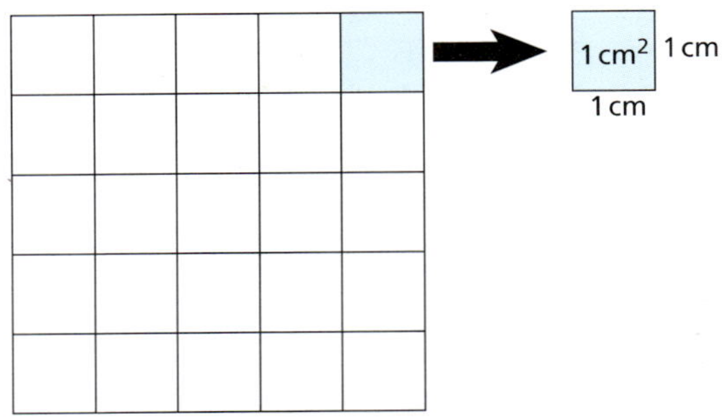

FIG 1.1.1 A square marked in centimetres

The area of a square or a rectangle can be found by placing it on a similar grid and counting how many squares it covers.

A square of sides 4 cm covers 16 squares on the grid. Its area must be 16 cm².

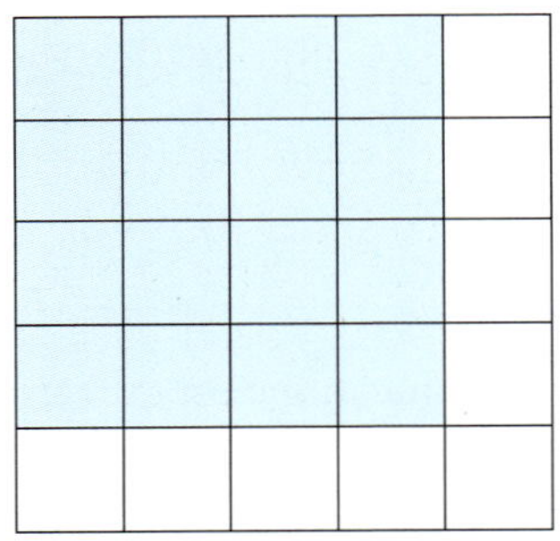

FIG 1.1.2 Area of a square of sides 4 cm

Composite shapes

The area of a composite shape can be found by cutting it into pieces and placing those pieces on a 1 cm² grid.

In Fig 1.1.3 the parts of the letter 'E' form a rectangle 5 cm long by 2 cm wide. The parts cover 10 × 1 cm², so the area of the letter must be 10 cm².

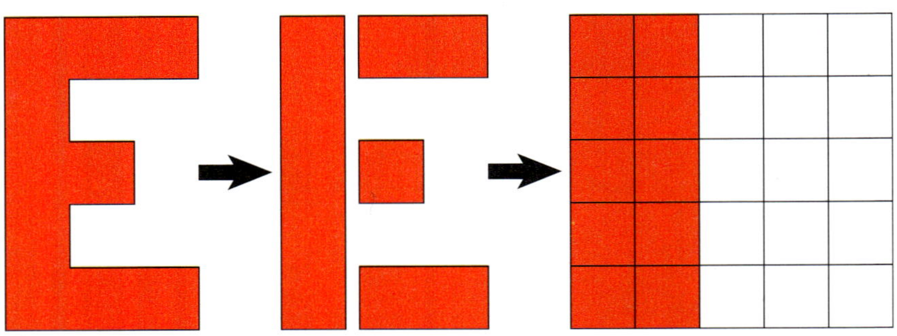

FIG 1.1.3 The parts of the shape cover 10 × 1 cm²

Finding the area of a shape by placing it on a grid works fine, but is of limited use. How could you use it to find the area of a large shape like a football pitch?

Formulae for calculating area

Table 1.1 shows information about the areas of the shapes opposite. Can you see a pattern between length, width and area?

The area of a square or a rectangle is equal to the product of the sides. In a square the length and width are equal, so:

area of square = $length^2$

The area of a rectangle is equal to the product of the length and the width:

area of rectangle = length × width

Notice that area is always measured in units squared. The unit of area depends on the unit used for the dimensions of the shape. We express small areas in cm^2 and larger areas in m^2.

Length	Width	Area
1 cm	1 cm	1 cm^2
4 cm	4 cm	16 cm^2
5 cm	3 cm	15 cm^2
5 cm	2 cm	10 cm^2

TABLE 1.1.1

Interesting fact

Very large areas, like the area of a country, are expressed in kilometres squared. The area of Jamaica is 10 991 km^2.

Key terms

area the region bound by the edges of a flat shape

centimetre squared (cm^2) and **metre squared (m^2)** units of area

Activity 1.1.1

Finding the area of surfaces in the classroom

You should work with another student on this activity. Make sure that each of you has an opportunity to measure and to record. Here is what you need:

- metre rule or measure tape.

Here is what you should do:

1. Measure the length and width of some square or rectangular surfaces in your classroom. These might be shapes like the top of your desk, the front of your textbook, the chalkboard, etc. Measure small surfaces in centimetres and large surfaces in metres.
2. Record your measurements in a table.
3. Calculate the area of each surface, including the unit.

Check your understanding

1. Remember to include the correct unit with each of your answers.

   a) Calculate the area of squares that have sides of the following length:

   i) 15 cm ii) 5 m iii) 2.5 cm.

   b) Calculate the area of rectangles that have sides of the following lengths:

   i) 18 cm and 12 cm iii) 5 m and 3.5 m.

2. A house brick has sides of 23 cm, 11 cm and 7.5 cm.

   Calculate the areas of the three sides of the brick you can see in the picture.

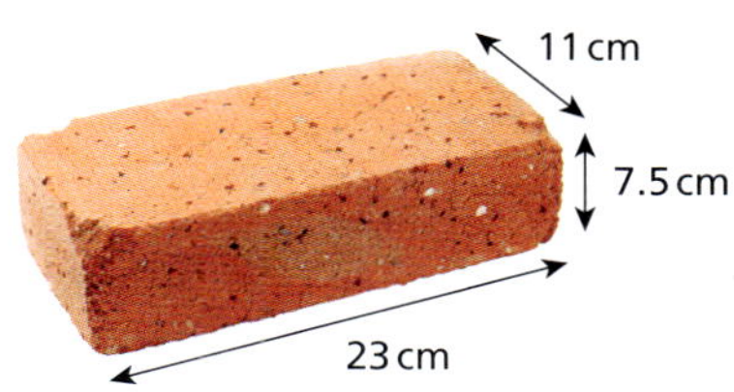

FIG 1.1.4

1.2 Area of irregular shapes

We are learning how to:

- use appropriate units for area
- estimate the area of irregular shapes

The areas of shapes can be compared by investigating whether shapes cover each other or not. Fig 1.2.1 shows a circle, a square and a triangle.

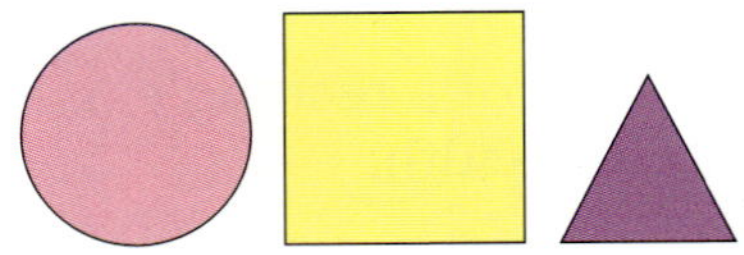

FIG 1.2.1 Three shapes

The triangle fits inside the circle and the circle fits inside the square.

So the areas of these shapes, from smallest to largest, is: triangle, circle, square.

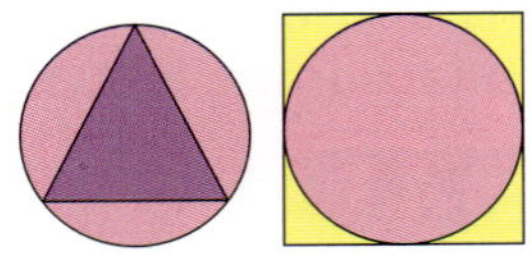

FIG 1.2.2 The triangle fits inside the circle, and the circle fits inside the square

Estimating the area

The area of a shape can be **estimated** by finding out how many squares of a 1 cm^2 grid it covers even when the sides are not whole numbers. The rectangle in Fig 1.2.3 has sides of 4.0 cm and 2.5 cm.

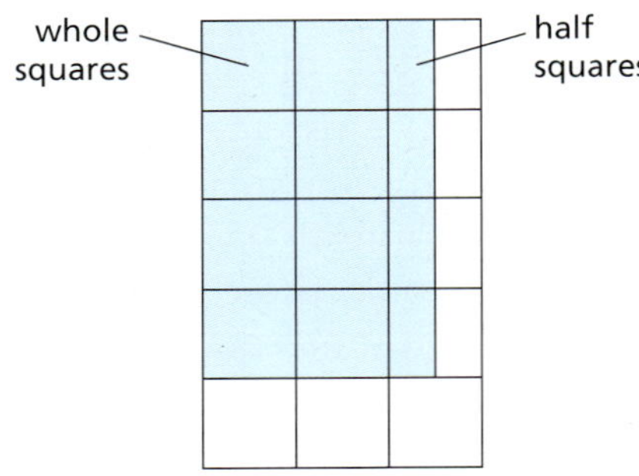

FIG 1.2.3 Area of a rectangle of sides 4.0 cm and 2.5 cm

The rectangle covers some whole squares and some half squares. There are:

- 8 squares which are all covered
- 4 squares which are half covered.

The estimated area of this rectangle is $8 + 4 \times 0.5 = 10.0$ cm^2. You can compare this with the value obtained using the formula given in the previous topic:

$$\text{area of rectangle} = 4.0 \times 2.5 = 10.0 \text{ cm}^2$$

When an area cannot be calculated using a formula, it has to be estimated. In Fig 1.2.4 the oval covers the squares of the grid by different amounts.

The area is estimated by counting up those squares that are at least half covered by the shape. This leads to an over-estimate of the area but this is compensated for by ignoring the squares that are less than half covered.

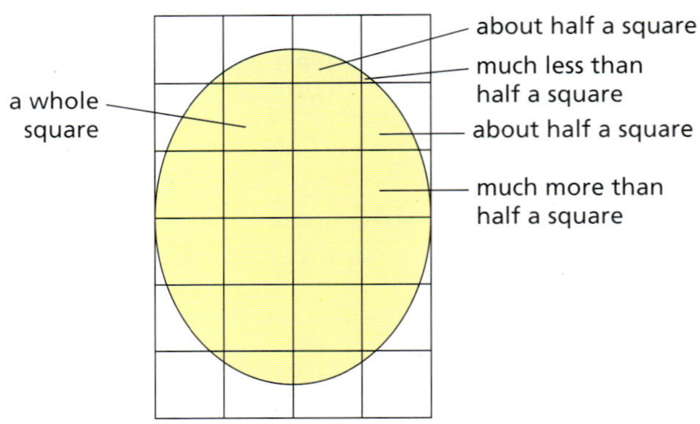

FIG 1.2.4 Area of an oval shape

Using this method, an estimate for the area of the oval is:

8 whole squares covered + 12 squares that are about half covered or more
= 20 squares

Since each square represents 1 cm^2, this represents an area of 20 cm^2.

Activity 1.2.1

Estimating the area of an irregular shape

Here is what you need:

- ruler 30 cm
- sheet of plain paper
- irregular shape cut in card.

Here is what you should do:

1. Draw a grid 10 cm by 10 cm on the plain paper using a pencil and ruler.
2. Place the irregular shape onto the grid and draw round it.
3. Count the number of squares that are at least half covered by the outline of the shape. You might find it helpful to shade each part as you add it, so that none are missed or counted twice.
4. Write down your estimate of the area of the shape.

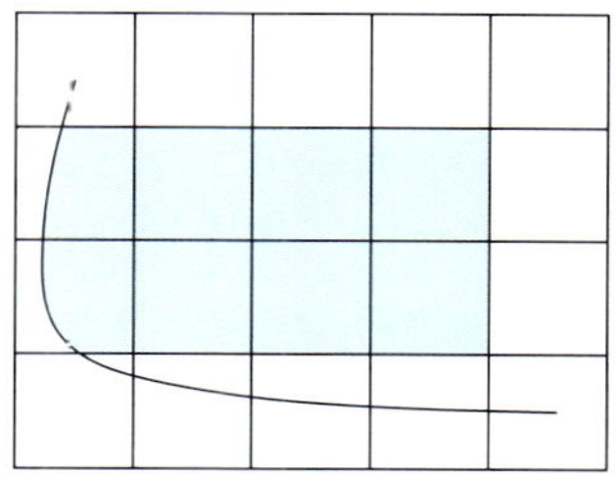

FIG 1.2.5 An irregular shape on a grid

Check your understanding

1. A student drew around an irregular shape that had been placed on a 1 cm^2 grid.

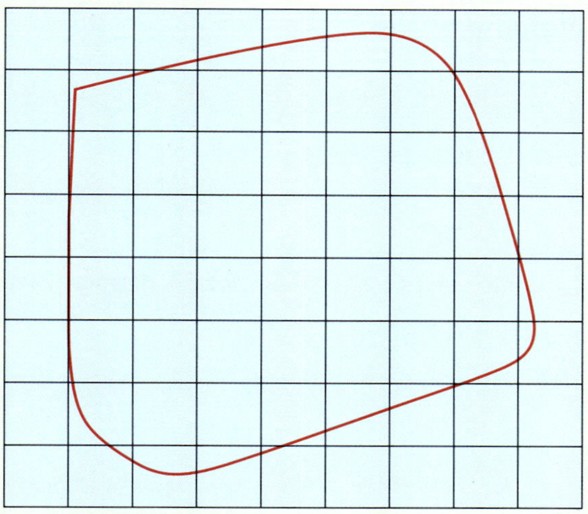

FIG 1.2.6 The student's irregular shape

Estimate the area of this shape using the method described above.

Key terms

irregular shape has sides and angles of different sizes or lacks symmetry

estimate an approximate amount

1.3 Volume of regular shaped objects

We are learning how to:

- formulate a simple working definition for the term 'volume'
- use appropriate units for volume
- calculate the volume of regular shaped objects

The **volume** of a solid shape is the amount of space that it occupies. A cube is a regular solid shape that has equal sides; that is, length = width = height.

When the sides of a cube are one centimetre, the cube occupies a space of one **centimetre cubed** or one **cubic centimetre**. This unit of volume is written as **cm**^{3}.

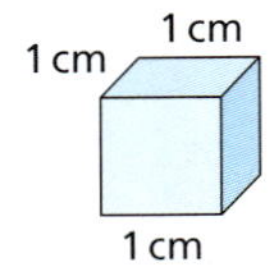

FIG 1.3.1 A cubic centimetre

Cubes and cuboids

Any cube can be broken down into smaller cubes with sides of 1 cm.

Fig 1.3.2 shows the number of cubes of sides 1 cm in some larger cubes. The volumes of the cubes are given in Table 1.3.1. Can you see a pattern between the length of the side of a cube and its volume?

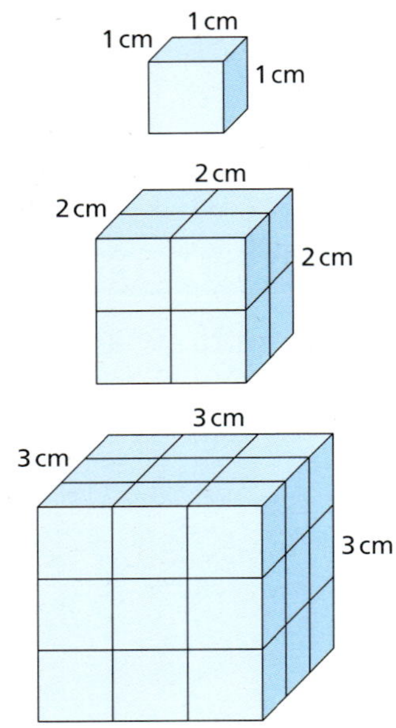

FIG 1.3.2 How many smaller cubes in each large cube?

Length of side of the cube	Number of cubes of sides 1 cm in the cube	Volume of the cube
1 cm	1	1 cm^3
2 cm	8	8 cm^3
3 cm	27	27 cm^3

TABLE 1.3.1

The volume of a cube is equal to: length × width × height.

The same formula can also be used to find the volume of a cuboid.

Cylinder and sphere

The volume of other regular solids can be found using suitable formulae.

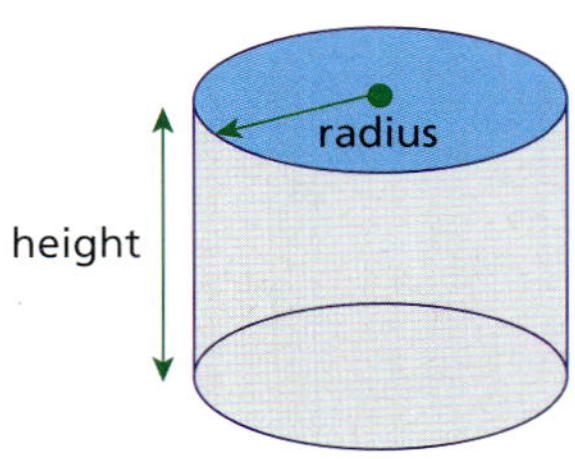

FIG 1.3.3 Cylinder

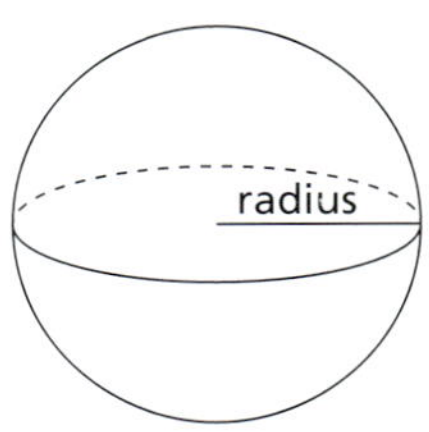

FIG 1.3.4 Sphere

Volume of a cylinder = base area × height = $\pi \times \text{radius}^2 \times \text{height}$

Volume of a sphere = $\frac{4}{3} \times \pi \times \text{radius}^3$

Composite shapes

The volume of a composite shape can be found by adding the volumes of the smaller shapes from which it is made.

The composite shape in Fig 1.3.5 can be separated into two cuboids.

The volume of the composite shape is the sum of the volume of the cuboids, which is:

$$7 \times 6 \times 2 + 6 \times 6 \times 4 = 84 + 144 = 228 \text{ cm}^3$$

The volume of large regular solid shapes is measured in exactly the same way but in **cubic metres**. A cubic metre, **m³**, is the volume of a cube that has sides of one metre.

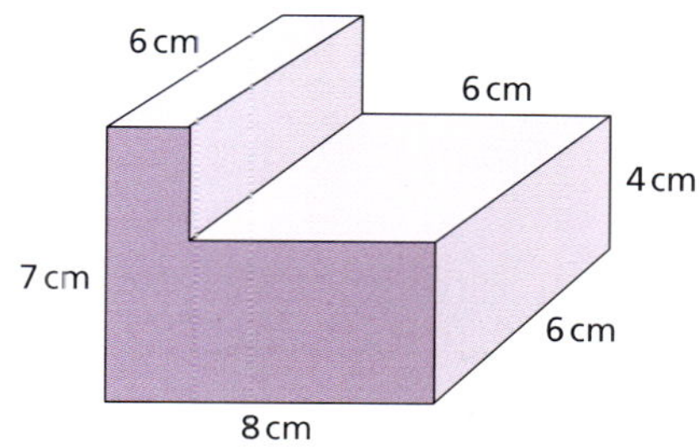

FIG 1.3.5 A composite solid shape

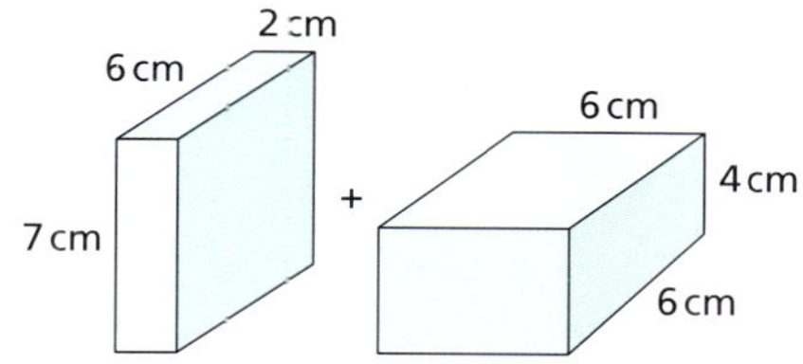

FIG 1.3.6 Composite shape separated into cuboids

Activity 1.3.1

Measuring regular solid shapes and calculating their volumes

You should work with another student on this activity. Make sure that each of you has an opportunity to measure and to record. Here is what you need:

- ruler 30 cm
- regular solids shapes labelled A, B, C, etc.

Here is what you should do:

1. Name the solid shape A.
2. Measure the dimensions of A and write them in a table.
3. Calculate the volume of the shape using a suitable formula.
4. Repeat this for all of the solid shapes you have been given

Key terms

volume the space that an object occupies

centimetre cubed/cubic centimetre (cm³) and **metre cubed/cubic metre (m³)** units of volume

Check your understanding

1. Count the number of cubes of sides 1 cm in each of the following shapes and use this to give their volume.

a)

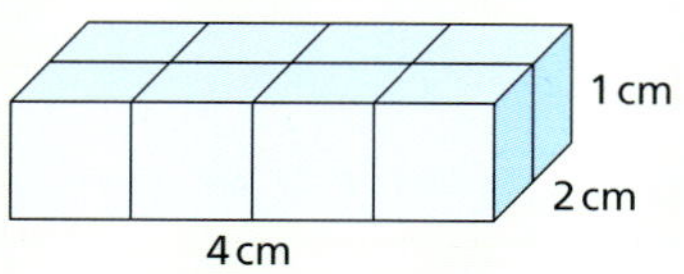

b)

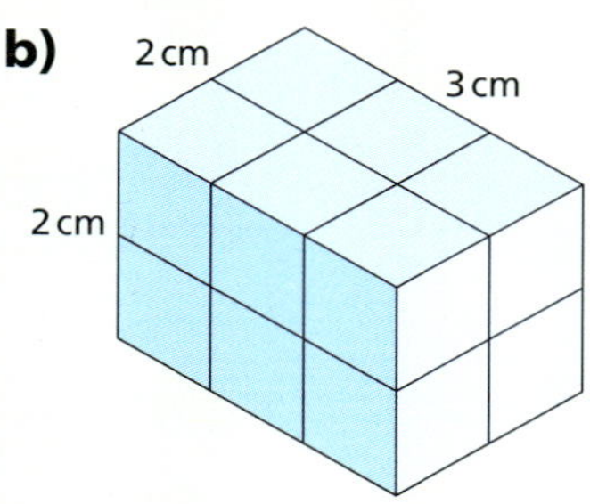

c)

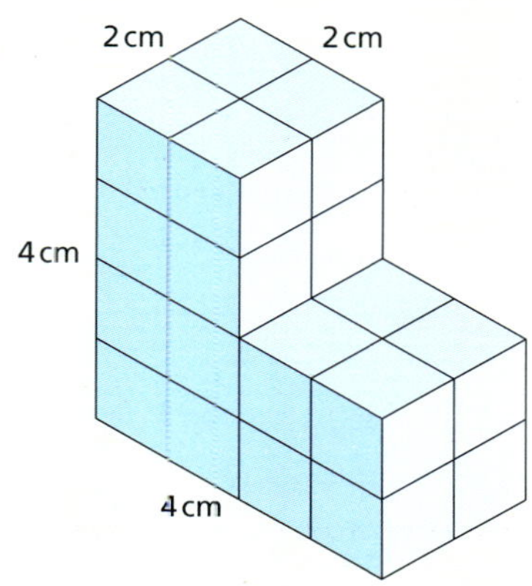

1.4 Volume of irregular shaped objects

We are learning how to:

- use appropriate units for volume
- determine the volume of irregular shaped objects

An **irregular solid** is one that doesn't have a normal shape, such as a cube, a cylinder or a sphere. It may have sides of different lengths and it lacks symmetry.

It isn't possible to divide an irregular shape into 1 cm^3 cubes, nor to apply a formula to find its volume.

FIG 1.4.1 An irregular solid

Finding volume by displacement

The volume of an irregular solid must be found by **displacement**.

When a solid object is submerged in a container of water, the level of the water rises. The object displaces a volume of water equal to itself. The new volume is equal to the sum of the volume of the water and the volume of the solid.

volume of shape = volume of water containing the shape – volume of water

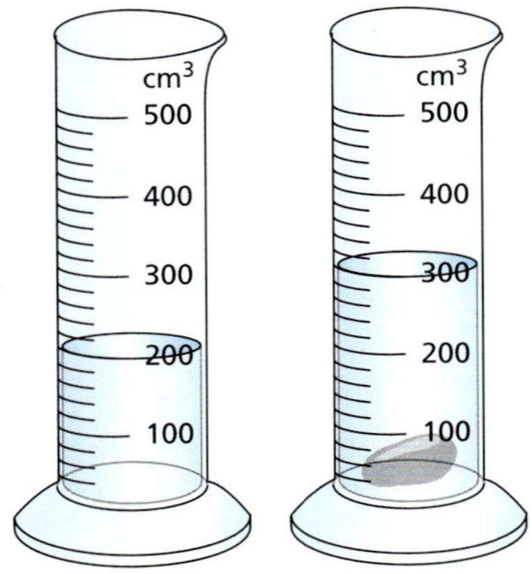

FIG 1.4.2 Measuring volume by displacement

Activity 1.4.1

Measuring the volume of an irregular solid by displacement

Here is what you need:

- irregular shaped objects
- measuring cylinder 250 cm^3
- string
- water.

Here is what you should do:

1. Half fill the measuring cylinder with water.
2. Tie a piece of string around object A. It needs to be long enough to lower the object into the measuring cylinder.
3. Record the initial volume of the water in the measuring cylinder in a table.
4. Gently lower the object into the water so it is totally submerged.
5. Record the final volume of the water.
6. Calculate the volume of the object.
7. Repeat steps 2 to 6 for all of the irregular solids you have been given.

Check your understanding

1. The following diagram shows the level of water in a measuring cylinder before and after introducing an irregular solid shape.

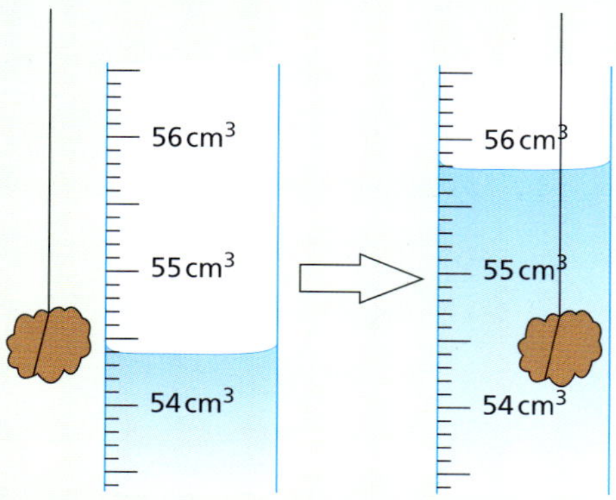

FIG 1.4.3 Levels of water in a measuring cylinder

a) What was:

i) the initial volume of water?

ii) the final volume of water?

iii) the volume of the irregular solid?

b) Why is it important that all of the shape is submerged in the water?

2. State whether the following shapes are regular or irregular.

a) 

FIG 1.4.4

b) 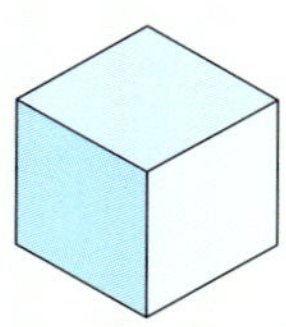

FIG 1.4.5

c) 

FIG 1.4.6

d) 

FIG 1.4.7

Key term

displacement a solid occupies the space that would be occupied by a liquid

Review of Working like a physicist

- Area is the space contained within the boundaries of a flat shape.
- Small areas are measured in centimetres squared (cm^2)while large areas are measured in metres squared (m^2).
- A square centimetre is the area bound by a square of sides 1 cm long.
- A square metre is the area bound by a square of sides 1 m long.
- The area of a square or a rectangle that has dimensions which are whole numbers can be found by placing the shape on a grid of 1 cm squares.
- The area of a regular square or rectangle can be calculated using the equation: area = length × width.
- The area of a composite flat shape can be found by calculating the area of each component.
- The relative areas of two or more shapes can be determined by finding whether the shapes can contain each other or not.
- The area of an irregular shape cannot be calculated but must be estimated.
- To estimate the area of an irregular shape, the shape is placed on a grid of 1 cm squares and only those squares that are at least half covered are counted.
- The volume of a solid shape is the amount of space it occupies.
- Small volumes are measured in centimetres cubed (cm^3)while large volumes are measured in metres cubed (m^3).
- A cubic centimetre is the volume of a cube of sides 1 cm long.
- A cubic metre is the volume by a cube of sides 1 m long.
- A cube, a cuboid, a cylinder and a sphere are examples of regular solid shapes.
- The volume of a cube or cuboid can be calculated using the formula: volume = length × width × height.
- The volume of a composite solid shape can be found by calculating the volume of each component.
- The volume of a cylinder can be calculated using the formula: volume = π × radius of base^2 × height.
- The volume of a sphere can be calculated using the formula: $\frac{4}{3} \times \pi \times \text{radius}^3$.
- The volume of an irregular solid shape cannot be calculated from a formula but must be found by displacement.

Review questions on Working like a physicist

1. A football pitch has sides of 105 m and 55 m.

   a) What shape is the football pitch?

   b) What area is covered by the football pitch?

2. In order to widen a road, a cuboid-shaped block of rock measuring 5 m by 8 m by 3.5 m high has to be removed.

   What volume of rock must be removed?

3. A cylindrical-shaped piece of wood has a radius of 10 cm and a height of 40 cm.

   a) Is this a regular or an irregular solid shape?

   b) What is the volume of this shape?

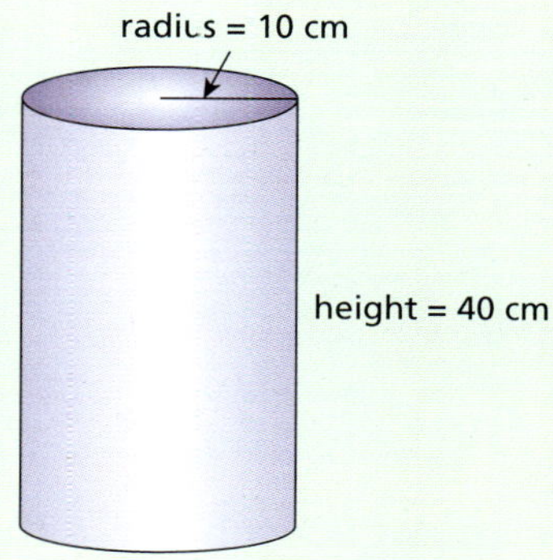

FIG 1.R.1

4. The following diagram shows a method of measuring volume by displacement.

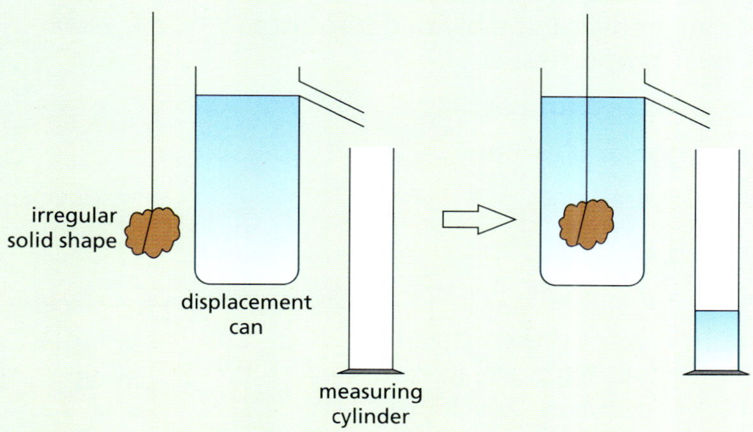

FIG 1.R.2

   a) Explain how this method works.

   b) Suggest how the method could be modified to find the volume of a piece of cork that floats on water.

Unit 2: Static electricity

2.1 Structure of the atom

We are learning how to:

- recall that atoms contain protons, neutrons and electrons and state their respective charges
- conduct investigations with due regard for safety
- work cooperatively in groups.

Atom comes from the Greek word *atomos*, which means indivisible. Atoms are the building blocks from which all things are made.

FIG 2.1.1 Some common elements: (a) aluminium, (b) carbon, (c) silicon

Interesting fact

The idea of a particle that could not be divided into anything smaller was first proposed over two thousand years ago by an Ancient Greek called Democritus.

An **element** is a substance that cannot be made into simpler substances. All of the atoms of one element are similar in structure, but different from the atoms of any other element.

There are 94 naturally occurring elements and many will be familiar to you. These elements can be divided into three groups according to their properties:

- metals, like aluminium and copper
- non-metals, like oxygen and carbon
- semi-metals, like silicon.

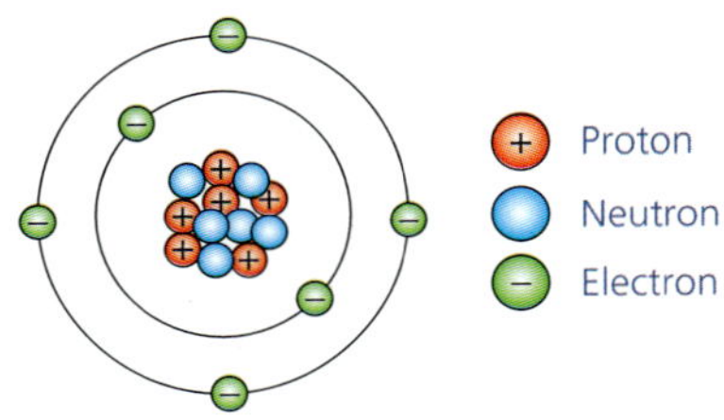

FIG 2.1.2 Structure of an atom

Sub-atomic particles

As knowledge increased, scientists discovered that atoms are not indivisible. They are, in fact, composed of even smaller particles, which are called **sub-atomic particles**.

An atom consists of a nucleus containing protons and neutrons. Surrounding the nucleus are layers of electrons. Table 2.1 summarises the properties of these three particles.

Notice that:

- The mass of an electron is only $^1/_{1836}$ the mass of a proton or a neutron, which isn't very much so it is usually written as 0.

Particle	Relative mass	Relative charge	Position in the atom
Proton	1	+1	In the nucleus
Neutron	1	0	In the nucleus
Electron	0 $\frac{1}{1836}$	−1	In shells around the nucleus

TABLE 2.1

- Although the masses of a proton and an electron are very different, the charges they carry are equal in size but opposite in charge.
- The neutron carries no charge at all.

Activity 2.1.1

Building a model of an atom of carbon

You should work in a small group for this activity.

Here is what you need:

- small polystyrene balls or similar spheres (12 large and six small)
- thin wire
- glue
- paints
- thread.

Here is what you should do:

1. Decide on colours to represent protons, neutrons and electrons.
2. Paint six of your balls with each colour. Remember that the electrons are smaller.
3. Glue your protons and neutrons together to form the nucleus of your atom.
4. Make a loop of wire a little larger than your nucleus and thread two of your electrons onto the wire. Keep them in position on your loop with glue. Remember the correct positioning of the electrons.

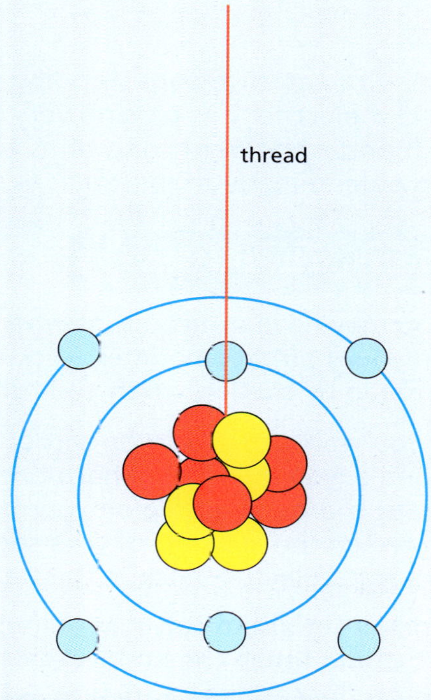

FIG 2.1.3 Model of a carbon atom

5. Make a second loop a little larger than the first; then thread the remaining four electrons onto it. Hold them in position with glue.
6. Attach a piece of thread with glue, firstly to the nucleus, and then loop it around the rings of electrons and hold it in position with glue.
7. Find a suitable place in your classroom to suspend your atom.

Check your understanding

1. Copy and complete the following sentences.

   a) Atom comes from a Greek word than means ....

   b) An .... is a substance that cannot be changed into simpler substances.

   c) An .... is the smallest particle of an element that can exist.

   d) .... particles are particles found inside an atom.

Key terms

atom the smallest particle of a substance that can take part in a chemical reaction

element substance that cannot be made into simpler substances

sub-atomic particles particles found within the atom

2.2 Investigating static electricity

We are learning how to:

- investigate the production of static electricity
- conduct investigations with due regard for safety.

The word **static** means standing still. We normally think about electricity as passing very quickly along metal wires, but under the right conditions electrical charge can build up on objects made of materials like glass and plastic.

FIG 2.2.1 Lightning is the discharge of static electricity to earth

Charging by friction

You may have noticed that when you use a plastic comb on your hair it sometimes crackles. Static electricity is being created by **friction** when you rub your hair with the comb.

If you walk across a nylon carpet and then touch a metal door handle, you may feel a sharp tingle at the end of your fingers. Static charge builds up on your body as a result of friction between the bottoms of your shoes and the carpet. The tingle is a small electric shock as the static electricity is discharged.

You might also receive an electric shock when you get out of a car and touch the metal door handle. Static charge builds up on the car due to friction as it moves through the air.

Under certain conditions charge even builds up in the atmosphere. Eventually the charge becomes so great it passes to earth in a great spark that we call **lightning**.

Interesting fact

The Ancient Greeks noticed that rubbing a material called amber on wool gives it unusual properties. It is able to pick up small objects likes hairs, threads, pieces of feathers and even dry leaves.

Charging with static electricity

Any insulator can be charged with static electricity by rubbing it with another insulator. In the laboratory, rods of glass or plastic are charged by rubbing them on fabrics such as wool or silk.

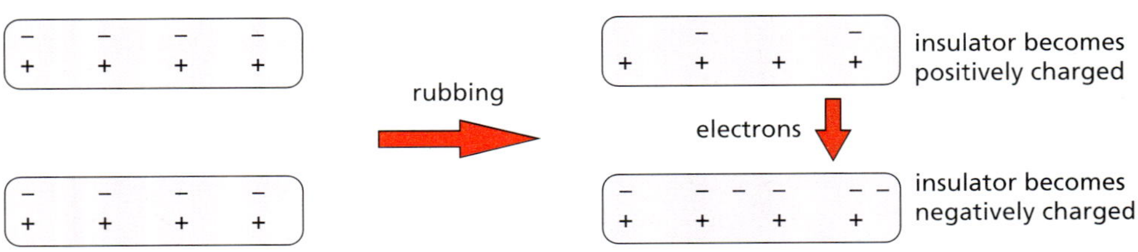

FIG 2.2.2 Charge is the result of the transfer of electrons

When the two insulators are rubbed together, electrons are transferred from one substance to the other. As electrons carry a negative charge, the insulator that loses electrons becomes positively charged while the insulator that gains electrons becomes negatively charged.

Charged objects interact with each other. If you bring two objects that have the same charge, either + or –, towards each other they will repel, but if you bring together two objects that carry opposite charges, that is + and –, they will attract each other.

Activity 2.2.1

Investigating some effects of static electricity

Here is what you need:

- plastic comb (must be clean and not greasy)
- glass rod
- rubber balloon
- piece of silk
- piece of wool
- small pieces of tissue paper.

Here is what you should do.

1. Rub the comb vigorously against your uniform for about five minutes, or until the comb feels warm.
2. Place the comb near some small pieces of paper.
3. Describe what you observe.
4. Rub a glass rod with a piece of silk.
5. Turn on a tap so there is a very thin column of water flowing.
6. Bring the charged glass rod close to the column of water but don't touch it.
7. Describe what you observe.
8. Blow up a balloon.
9. Rub the balloon on a piece of wool or a woollen jumper.
10. Place the rubbed part of the balloon against the wall and then carefully move your hands away.
11. Describe what you observe.

Key terms

static not moving

friction force that opposes motion

lightning movement of electrical charge between the sky and the ground

Check your understanding

1. A student rubbed an inflated balloon on her woollen jumper and then placed it on a wall. When she took her hands away, the balloon stayed in the same position.

   a) Explain why the balloon becomes negatively charged when it is rubbed on the woollen jumper.

   b) With the help of a diagram, explain why the balloon remains 'fixed' to the wall and does not fall.

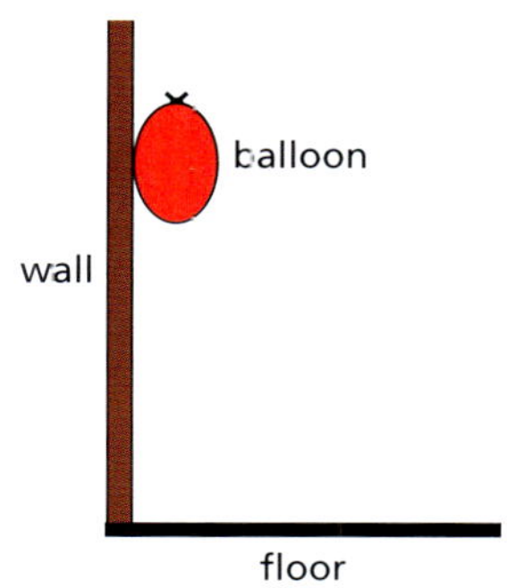

FIG 2.2.3 The student's experiment

Grade 9

2.3 Uses and hazards of static electricity

We are learning how to:

- describe useful applications and hazards of static electricity.

Electrostatic charging has a number of important applications, but the build-up of electrostatic charge on insulating materials not connected to earth also creates some serious hazards.

Uses of electrostatic charge

Photocopiers

Electrostatic charge provides a means of copying documents.

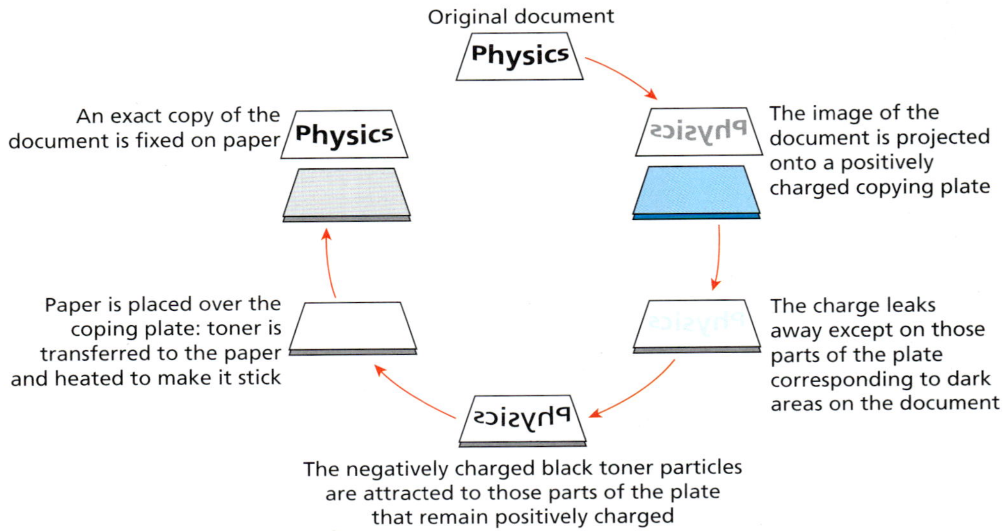

FIG 2.3.1 Photocopying

When a document is scanned on a photocopier, the image is projected onto a positively charged plate. Where light falls on the plate, the charge leaks away. This leaves a charged area on the plate corresponding to the dark part of the document.

Negatively charged black (or colour) toner powder is then attracted to those parts of the plate that are still charged. This produces a mirror image of the document on the plate.

Paper is placed over the plate. The toner is transferred to the paper and fixed by heating. The result is a perfect copy of the original document.

Problems caused by electrostatic charge

Lightning

Under certain weather conditions, massive amounts of electrostatic charge may build up in the atmosphere and finally discharge to earth as lightning. Lightning tends to discharge through objects sticking up above the ground such as trees, tall chimneys and towers. That is why standing near these objects in stormy weather is dangerous. **Lightning conductors** are attached to many buildings to prevent or minimise damage in the event of a lightning strike. The conductor consists of thick copper or brass wire that provides the electric charge with an easy passage to earth.

Polythene manufacture

During the manufacturing process, sheets of polythene (and other plastics) pass around various rollers, as a result of which an electrostatic charge builds up on the polythene. The electrostatic charge is removed by placing a bar containing a **radioactive isotope** across the polythene, just above but not touching it. The isotope causes the air above the polythene to ionise, and the charge on the polythene is neutralised.

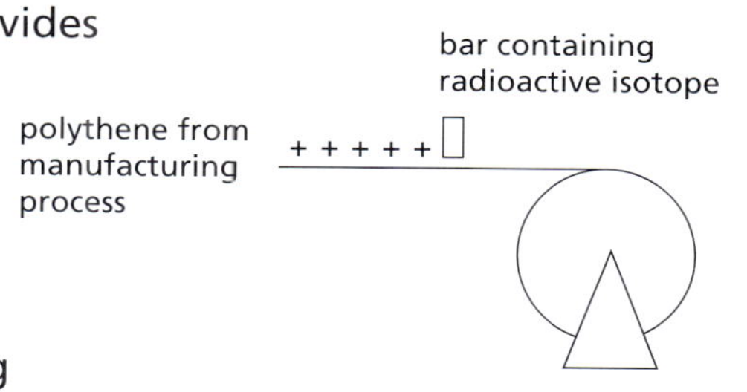

FIG 2.3.3 Static electricity builds up on polythene during manufacture

Activity 2.3.1

Assessing risk when transferring fuels

The movement of liquids along pipes results in the build-up of electrostatic charge. This is a particular danger with petrol, diesel or aviation fuel.

Consider the problems created by static electricity when transferring fuels:

- How might electrostatic charge be built up?
- What hazards result from the build-up of electrostatic charge?
- What safety procedures should be used?

Key terms

electrostatic static electrical charge

lightning conductor metal strip to protect from damage due to lightning

radioactive isotope substance that emits radiation

Check your understanding

1. During the manufacturing process, artificial fertiliser powder acquires an electrostatic charge. Fig 2.3.5 shows what happens when bags full of fertiliser powder are sealed.

   a) How does the polythene bag become charged?

   b) Why does the electrostatic charge make it difficult to close the bags?

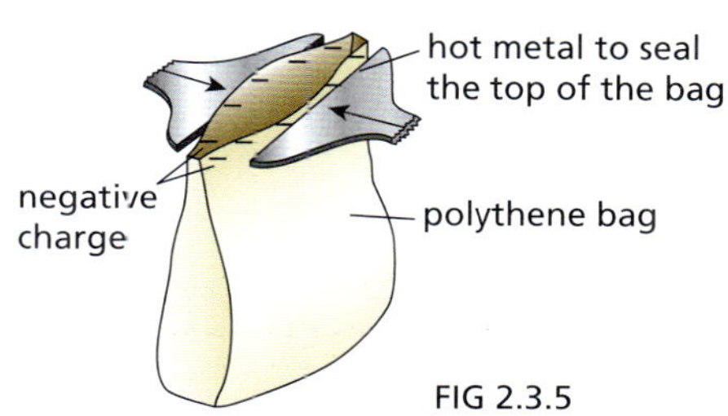

FIG 2.3.5

Review of Static electricity

- Atoms are the basic building blocks from which all things are made.
- An element is a substance that cannot be made into simpler substances.
- Elements may be metals, non-metals or semi-metals.
- An atom is the smallest particle of an element that can exist.
- An atom contains the sub-atomic particles: protons, neutrons and electrons.
- Protons and neutrons make up the nucleus of an atom; electrons are found in shells outside the nucleus.
- Protons and neutrons have a relative mass of 1; electrons have a relative mass of 0.
- Protons and electrons have equal but opposite charges.
- Static electricity is electrical charge built up on an insulator.
- Insulators can be charged by friction.
- Electronic charge is the result of electrons being transferred between insulators.
- Insulators that lose electrons become positively charged; insulators that gain electrons become negatively charged.
- Like charges repel while unlike charges attract.
- Electrostatic charge sometimes makes materials behave in unusual ways.
- Static electricity has important uses, such as photocopying.
- Static electricity creates some hazards including: damage by lightning strikes and build-up of charge during polythene manufacture.

Review questions on Static electricity

1. Draw a diagram of an atom that contains two protons, two neutrons and two electrons. Choose appropriate symbols for these particles and include a key alongside your drawing.

2. **a)** What does the word 'static' mean?

b) Why is there sometimes lightning during a storm?

c) Why does static electricity build up on a car body?

3. Fig 2.R.1 shows details of a device that refreshes the air in a room by removing particles of dust.

Explain how this device works.

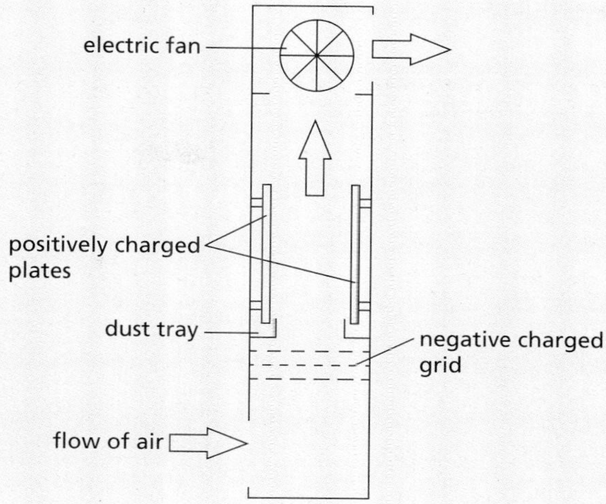

FIG 2.R.1

4. Explain each of the following observations.

a) When a person combs their hair with a plastic comb, the comb becomes charged but this does not happen when they use an aluminium comb.

b) Clothing made of synthetic fabrics sometimes crackles when pulled apart after drying in a tumble drier.

c) A rescue person hanging from a helicopter by a rope carries an electrostatic charge.

d) The screen of a television becomes dusty even when the rest of the television is clean.

5. What force is created when the following are brought together:

a) two positive charges?

b) two negative charges?

c) a positive and a negative charge?

6. Diamonds are electrical insulators. They are found in dirt, which does conduct electricity. Fig 2.R.2 shows an electrostatic method for separating diamonds from dirt.

a) Explain how the method works.

b) The particles are closely spaced at W on the conveyor belt. Explain why they are further apart at X.

c) Explain the curved path of the diamonds between Y and Z.

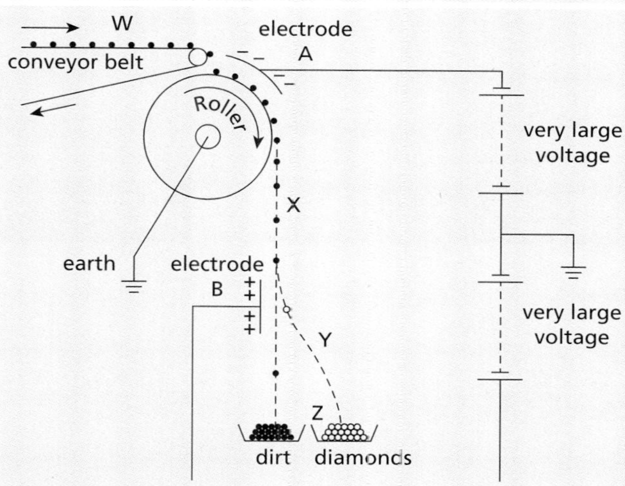

FIG 2.R.2

Unit 3: Current electricity

3.1 Conductors and insulators

We are learning how to:

- distinguish between electrical insulators and conductors
- relate flow of current to conduction

Electricity is a form of energy. It can pass easily through some materials but not others.

- Materials that allow an electric current to flow through them are called electrical **conductors**.
- Materials that prevent the flow of an electric current are called electrical **insulators**.

Activity 3.1.1

Conductors and insulators

Here is what you will need:

- battery of three cells
- lamp
- two crocodile clips
- connecting wires
- samples of materials – for example, aluminium, copper, plastic, rubber, wood.

Here is what you should do:

1. Connect the components of the circuit together as shown in Fig 3.1.1.

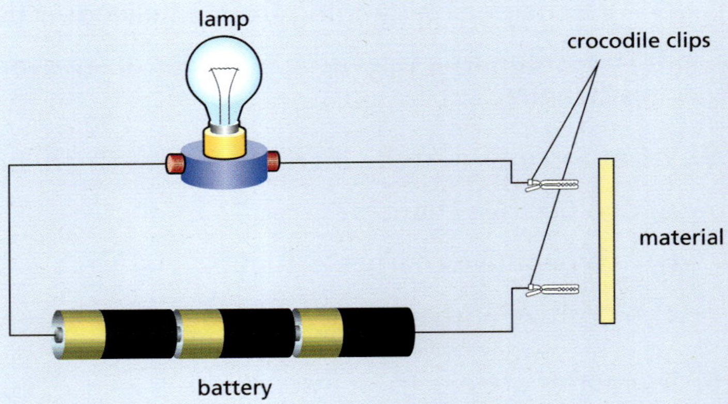

FIG 3.1.1

2. Before testing the sample materials, test the circuit by touching the crocodile clips together. If the circuit is complete the lamp should light up. If the lamp does not light up, check all the connections.
3. Take the first sample of material and clip the crocodile clips to each end of it. If the lamp lights the material is a conductor. If the bulb does not light the material is an insulator.
4. Present your observations in the form of a table. On one side write the names of the conducting materials, and on the other write the names of the insulating materials.

Metallic structure

Metals are all excellent conductors of electricity. To understand why, we need to consider **metallic structure**. Metals consist of a matrix of particles surrounded by a 'sea' of negatively charged **electrons**. These electrons are delocalised and free to move about.

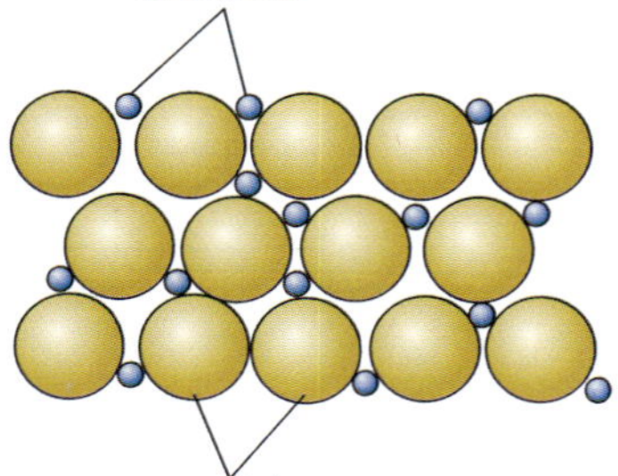

FIG 3.1.2 Structure of a metal

When a conductor, such as a metal wire, is connected to a battery, electrons flow through the conductor carrying electrical charge. Electricity is a flow of electrons.

Insulators, such as plastics and glass, do not have delocalised electrons so they are unable to conduct electricity.

Check your understanding

1. Arrange the following materials into two lists: conductors and insulators.

aluminium	copper	glass	iron
plastic	rubber	steel	wood

Interesting fact

The best conductor is the metal silver, followed by copper, then gold and then aluminium. Why are electric wires made of copper and not silver?

Key terms

conductors materials that allow an electric current to flow through them

insulators materials that prevent the flow of an electric current

metallic structure a matrix of particles surrounded by a 'sea' of negatively charged electrons

electrons negatively charged particles

3.2 Electricity and safety

We are learning how to:

- distinguish between electrical insulators and conductors
- use electricity safely

Electricity is potentially dangerous so we must learn how to use it safely.

Some plugs have two pins and some have three. Sockets may have caps to prevent children poking things into them.

FIG 3.2.1 An electrical appliance is connected to the mains electricity supply by a plug, which fits into a socket

Activity 3.2.1

Examining a plug and socket

Here is what you need:

- plug
- socket.

SAFETY

Even though plugs and sockets are made of plastic, they should never be touched with wet hands. There is a danger that wetness will pass into the plug or socket and this can result in an electric shock.

Here is what you should do:

1. Look carefully at the plug and observe where the wires from the appliance will be connected.
2. Notice that when the plug is placed fully in the socket there are no exposed metal parts.
3. Look carefully at the socket and where the wires from the supply will be connected.
4. Notice that the holes for the pins of the plug are very small, which prevents people accidentally pushing something into the holes.
5. Notice also that the plug pins must be fully into the socket before they connect with the electricity supply. This means that no parts of the plug pins are exposed.
6. From what type of material are the bodies of the plug and socket made?

Wires in an electrical circuit

Two wires are needed to make a circuit. Some plugs have three pins and three wires. The third wire is called the **earth wire**. It protects the user should a fault develop in the appliance.

If the wires in a metal table lamp came loose and touched the metal body, anyone touching it would receive an electric shock. The earth wire is connected to the metal body. Most of the current will pass through it so the user would get a small shock and not come to any harm.

An electrical appliance that has a plastic body does not need to be earthed because plastic is an insulator. Such appliances are described as **double insulated**.

FIG 3.2.2 Metal-bodied appliances should be earthed

FIG 3.2.3 Plastic-bodied appliances do not need to be earthed

Check your understanding

1.

FIG 3.2.4 Gabriella's hairdryer

a) How does Gabriella connect her hairdryer to the power supply?

b) Why should Gabriella not use her hairdryer when her hands are wet?

c) What is the outer casing made of?

d) Will Gabriella's hairdryer have a 2-pin or a 3-pin plug? Explain your answer.

Interesting fact

Electricians use a detector to locate electrical wiring in walls and floors to avoid accidentally drilling through a cable and getting an electric shock.

Key terms

earth wire wire in a plug that is connected to the metal body of the appliance to protect the user in the event of a fault

double insulated an appliance that has a body made of an insulating material

3.3 Complete circuit

We are learning how to:

- construct simple electrical circuits
- identify a complete circuit.

We use mains electricity to power most appliances in our homes but it is far too dangerous for building circuits in the laboratory. Instead we use a **cell** or a **battery**. They provide much less electrical energy than the mains supply.

You might use the words 'cell' and 'battery' to mean the same thing in everyday language but in science these terms have particular meanings. A cell is what is often incorrectly called a battery, and a battery is a combination of two or more cells.

FIG 3.3.1 This is a single cell

FIG 3.3.2 When two or more cells are used together to power an electrical device, they are called a battery

Activity 3.3.1

Building simple circuits

Here is what you need:

- cell in holder
- lamp
- connecting wires.

Here is what you should do:

1. Make the circuit shown in Fig 3.3.3.

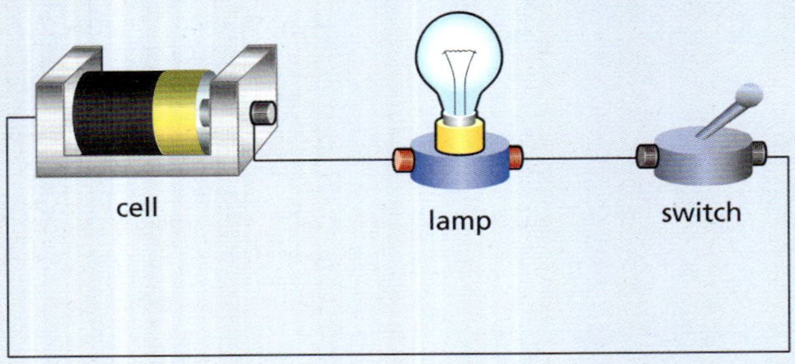

FIG 3.3.3

2. Draw the circuit and, alongside, say whether the lamp lit up or not when the switch was closed.
3. Make up five more different circuits.
4. Make a drawing of each circuit and say whether the lamp lights or not when the switch is closed.
5. Look at the circuits you made in which the lamp lit up. Can you see anything similar about them?

Current direction in an electrical circuit

A **circuit** is a complete pathway around which an electric current can flow. The pathway must be made of a conductor such as a metal wire.

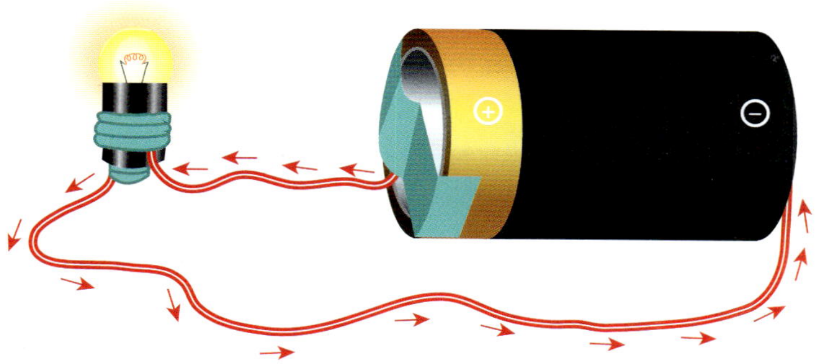

FIG 3.3.4 Direction of conventional current

A cell has a positive (+) terminal and a negative (–) terminal. Conventionally, the direction of the electric current is taken to flow from the positive terminal to the negative terminal. This is called **conventional current flow.**

In reality, the current is the result of a flow of electrons. Since electrons carry a negative charge, they actually flow from the negative terminal (where there are a lot of negative charges) to the positive terminal of the cell (where there are fewer negative charges). This is called **electron flow** and is in the opposite direction to the conventional current.

Interesting fact

You can create electricity using fruit and rods of two different metals, such as zinc and copper. Juicy fruits such as limes, oranges or lemons work best.

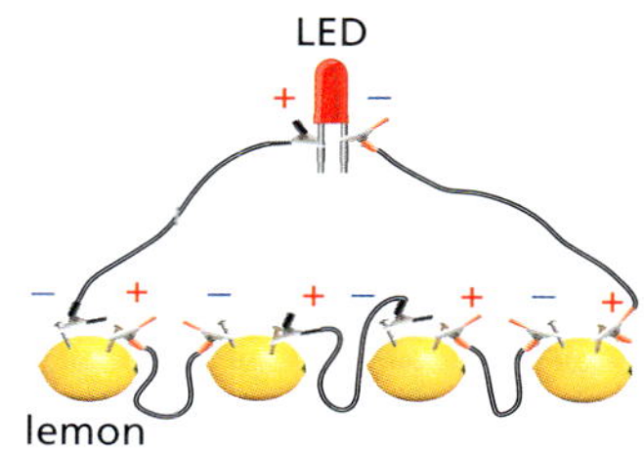

FIG 3.3.5 A fruit battery

Each lemon is a cell. When several are connected together, they produce enough electricity to light a small device called a light emitting diode (LED).

Check your understanding

1.

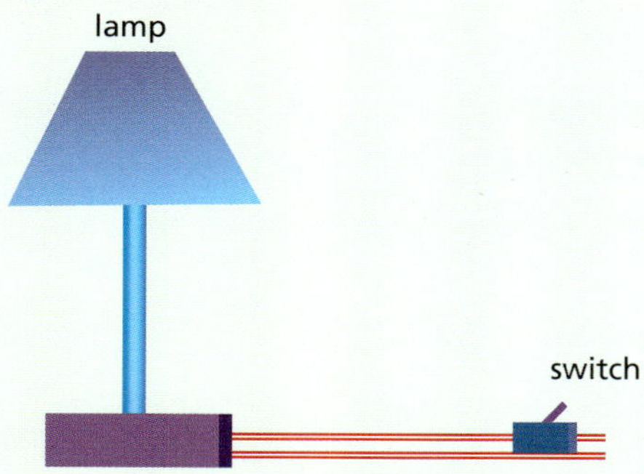

FIG 3.3.6 A lamp connected to the mains electricity supply

Explain, in terms of flow of current, how the switch is able to control the current in the lamp circuit in Figure 3.3.6.

Key terms

cell electrical power source often incorrectly called a battery

battery a combination of two or more cells

circuit a complete pathway around which an electric current can flow

conventional current flow current taken to flow from the positive terminal to the negative terminal of a cell

electron flow electrons flow from the negative terminal to the positive terminal of a cell

3.4 Cells and lamps

We are learning how to:

- construct simple electrical circuits
- observe the effect of using different numbers of cells and lamps in a circuit.

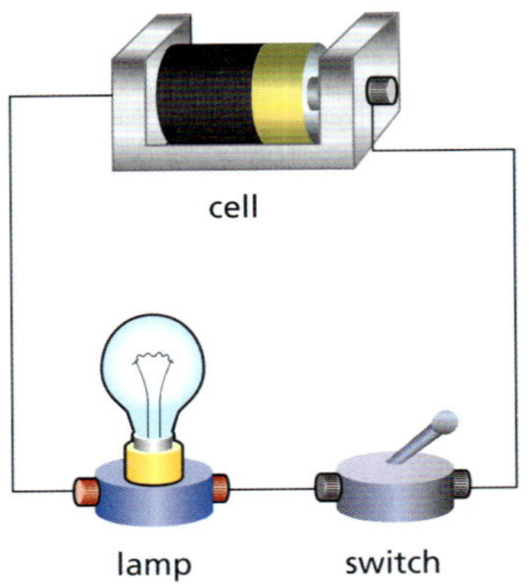

FIG 3.4.1 A simple lighting circuit might consist of a cell, a lamp and a switch

When additional components are added to a circuit, they can often be added in different ways, with different results.

Activity 3.4.1

Making circuits with cells, lamps and switches

Here is what you need:

- two cells
- two lamps
- switch
- connecting wires.

Here is what you should do:

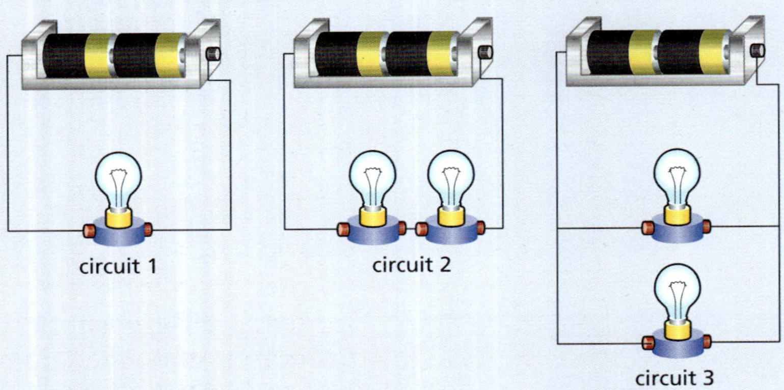

FIG 3.4.2

1. Build circuit 1.
2. Reverse the direction of one of the cells in circuit 1. Does the lamp still light up?

3. Build circuit 2 and make a note of how brightly the lamps glow.
4. Build circuit 3.
5. Do the lamps in circuit 3 glow less brightly, as brightly or more brightly than in circuit 2?
6. Include a switch at some different places in circuit 3. Do the lamps turn on and off differently according to the position of the switch?

Potential difference

There is a difference in potential energy (**potential difference**) between the terminals of a cell. This is measured in **volts (V)**. A single cell has a potential difference of 1.5 V, which is usually written on the side of it.

a) 1.5 V + 1.5 V = 3.0 V   b) 1.5 V – 1.5 V = 0.0 V

FIG 3.4.3 **a)** When two cells are arranged so that their terminals point in the same direction, the overall potential difference of the battery is the sum of the potential differences of the cells, i.e. 3.0 V

b) When cells are arranged so that their terminals point in opposite directions, the overall potential difference of the battery is the difference between the potential differences of the cells, i.e. 0 V

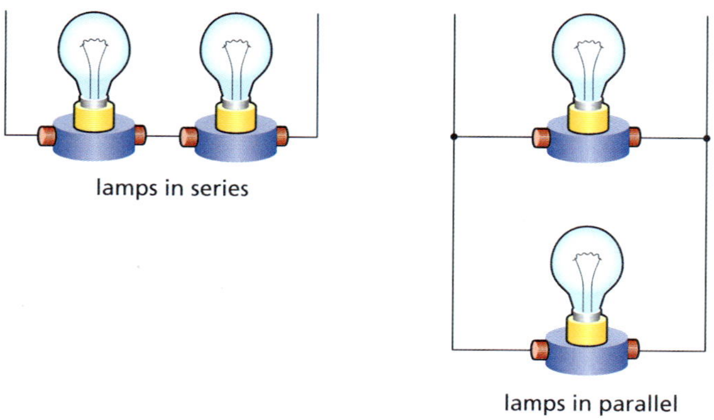

FIG 3.4.4 Two lamps can be connected in a circuit in two different ways

Check your understanding

1. Draw a circuit containing two cells, two lamps and one switch so that both lamps are on when the switch is closed and one lamp remains on when the switch is open.

Interesting fact

A car battery really is a battery in the scientific sense of the word.

It consists of six cells joined together inside a tough rubber casing. The potential difference of each cell is 2.0 V so the potential difference of the battery is 2.0 V added six times, i.e. 6 × 2 = 12.0 V.

Key terms

potential difference difference in potential energy between two points

volts (V) unit of potential difference

3.5 Measuring current

We are learning how to:

- construct simple electrical circuits
- measure current using an ammeter.

The amount of current passing in a circuit is measured using an **ammeter**. The unit of current is the **ampere (amp, symbol A)**.

In the activities we carry out in the laboratory, the current is often less than 1 ampere so we measure current in milliamperes or mA. There are 1000 milliamperes in 1 ampere.

$$1\ \text{A} = 1000\ \text{mA}$$

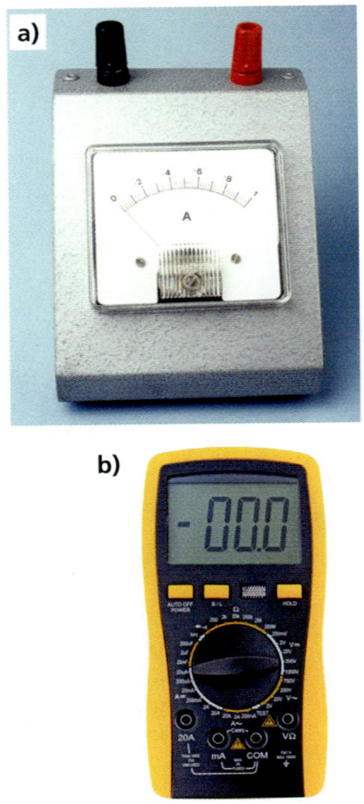

FIG 3.5.1 **a)** Analogue ammeter **b)** Digital ammeter

Ammeters

Electric current is measured by an ammeter. There are two types in common use.

An analogue ammeter has a moving pointer and the current is read from a scale where the pointer stops. A digital ammeter gives a direct numerical readout. It may be part of a **multimeter**, which has many different uses.

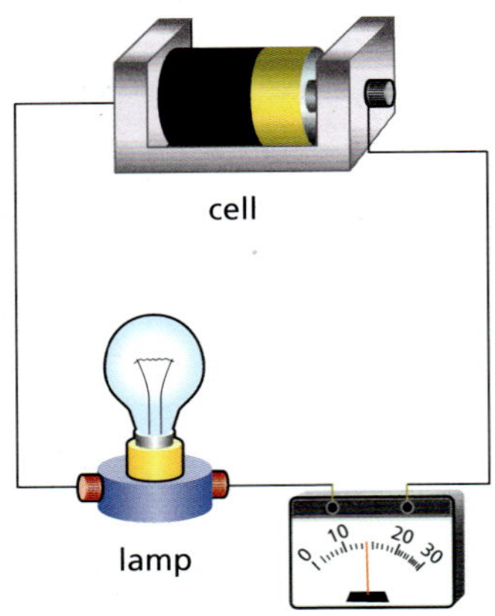

FIG 3.5.2 An ammeter is always connected in series to the component whose current it is measuring

Activity 3.5.1

Measuring the current at different points in a circuit

Here is what you need:

- battery containing two cells
- three lamps
- ammeter
- connecting wires
- long length of nichrome wire
- crocodile clip.

Here is what you should do:

1. Build the circuit shown in Fig 3.5.3.

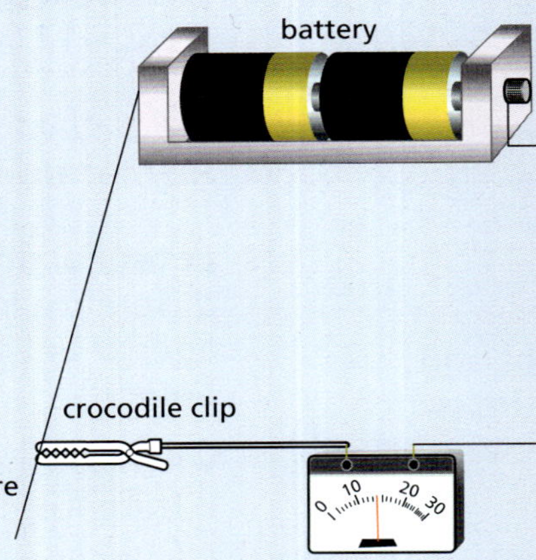

FIG 3.5.3

2. Connect the ammeter to different points on the nichrome wire using a crocodile clip.
3. How does the reading on the ammeter change with changes in the length of the nichrome wire in the circuit?
4. Build the circuit shown in Fig 3.5.4.

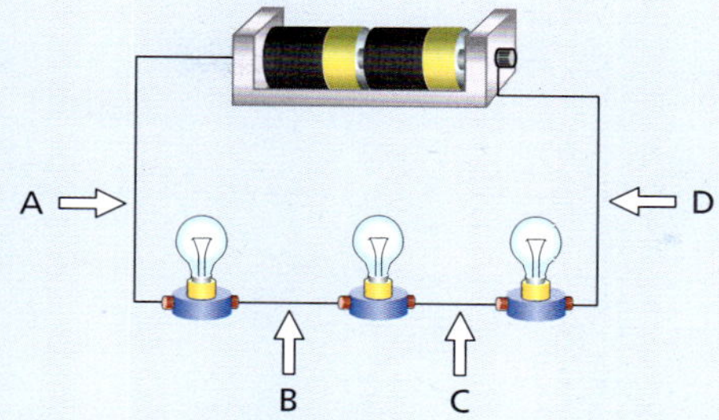

FIG 3.5.4

5. Connect the ammeter at point A in the circuit and record the current.
6. Repeat this at points B, C and D, recording the current each time.
7. What can you say about the current at different points in a circuit?

Nichrome wire resists the flow of current in a circuit. The greater the length of nichrome wire included in a circuit, the smaller the current that flows through it.

When components such as lamps are connected in series in a circuit, the current is the same no matter where in the circuit it is measured.

Check your understanding

1. a) What is the reading on this milliammeter:
   i) in milliamperes?
   ii) in amperes?

   b) Draw a diagram to show how an ammeter should be placed in a circuit containing a battery of two cells and one lamp, so that the current flowing through the lamp can be measured.

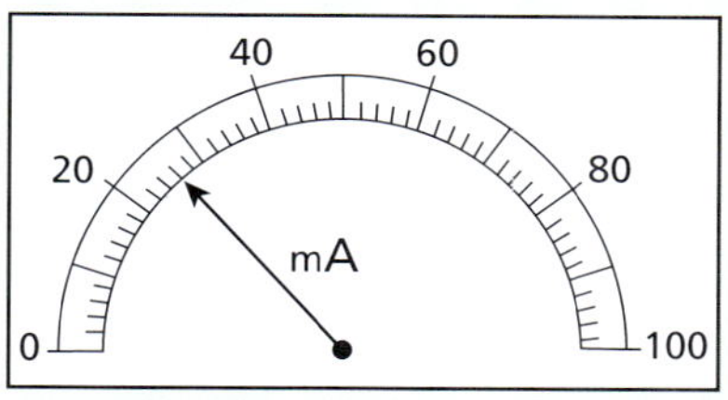

FIG 3.5.5 The reading on a milliammeter

Interesting fact

Electrical appliances are rated according to the amount of current that flows through them. Appliances that have small currents include lights, computers and televisions. Appliances that have high currents are those that supply heat in some way, such as irons, kettles and ovens.

Key terms

ammeter instrument used to measure current in a circuit

ampere (amp, symbol A) unit of current

multimeter instrument used to measure current and other properties of a circuit

3.6 Circuit symbols

We are learning how to:

- represent simple circuits using diagrams
- identify components of an electrical circuit from their symbols.

It would be possible to draw all of the **components** in an electrical circuit diagram but this would be time-consuming. It is much easier to draw the circuit using **symbols** to represent each of the components.

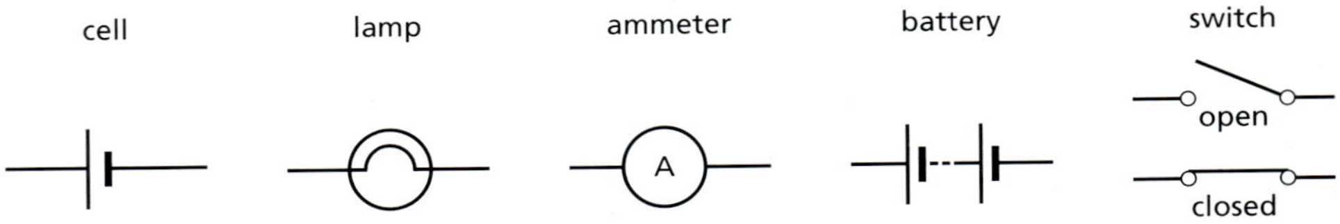

FIG 3.6.1 These symbols for components are used and understood by scientists all over the world

Notice that the symbol for a cell has two vertical lines. The long, thin line represents the positive (+) terminal and the shorter, thicker line represents the negative (–) terminal. It is important that you draw the symbol in the correct direction in a circuit diagram.

Circuit diagrams

To draw a **circuit diagram**, we draw the appropriate symbols and then connect them together by lines to represent connecting wires.

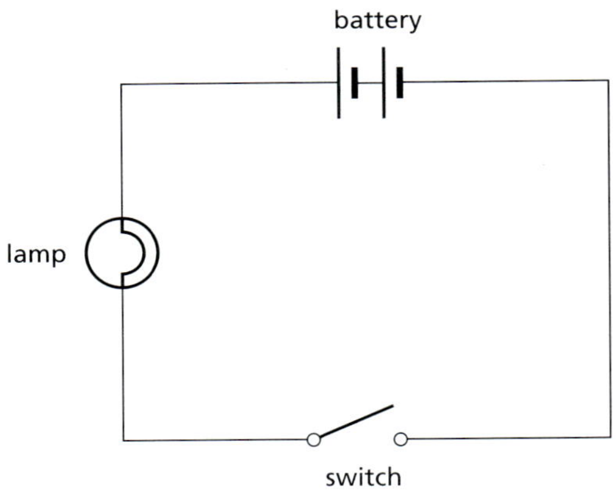

FIG 3.6.2 A circuit diagram for a circuit containing a battery, a lamp and a switch all connected in series

Notice that if a battery is composed of several cells, repeating the symbol for a cell can be tedious. It is much easier to join the symbols for two cells by a dashed line.

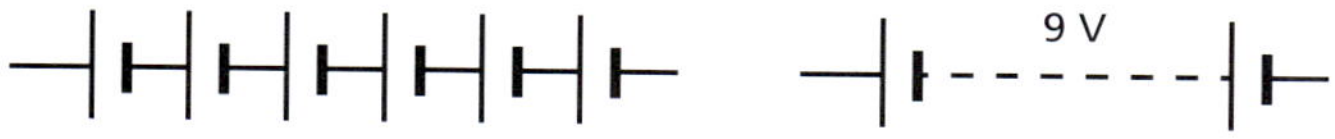

FIG 3.6.3 How to represent a battery composed of six cells

Activity 3.6.1

Drawing circuit diagrams

You will not need any equipment or materials for this activity.

Here is what you should do:

1. Look back at Figs 3.5.2, 3.5.3 and 3.5.4 and redraw the circuits using suitable symbols.
2. Would you say that it is easier to draw circuits using symbols than drawing the components?
3. Would you say that circuits drawn in symbols are easier to understand than circuits in which the components are drawn?

Check your understanding

1. Name the components in this electrical circuit.

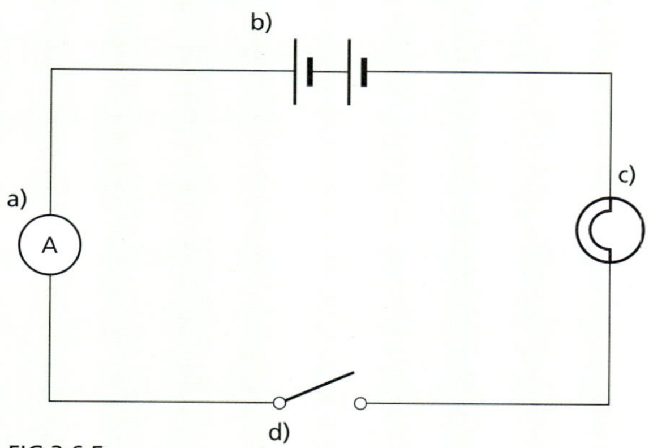

FIG 3.6.5

2. Draw a circuit diagram containing a battery of three cells, two lamps and a switch connected in series.

Interesting fact

There are two symbols commonly used to represent a lamp in circuit diagrams.

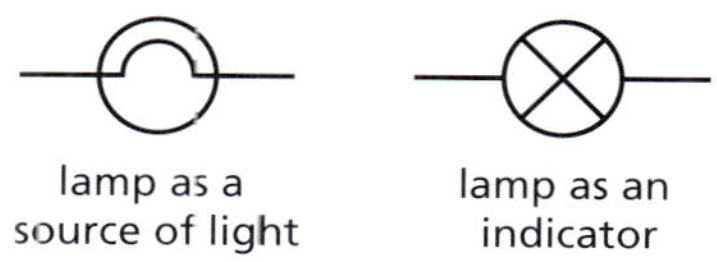

FIG 3.6.4 Symbols for a lamp

One symbol is used when the lamp is a source of light, such as in a torch circuit. The other is used when the lamp is an indicator of some kind, such as a light that comes on when an appliance is in use.

Key terms

components parts of a circuit

symbols signs used to represent something

circuit diagram diagram showing how components in a circuit are connected

3.7 Constructing circuits from circuit diagrams

We are learning how to:

- represent simple circuits using diagrams
- construct a circuit from a circuit diagram.

Drawing circuit diagrams

To build a circuit, we need to examine a circuit diagram in order to:

- identify the electrical components
- determine how the components are connected together.

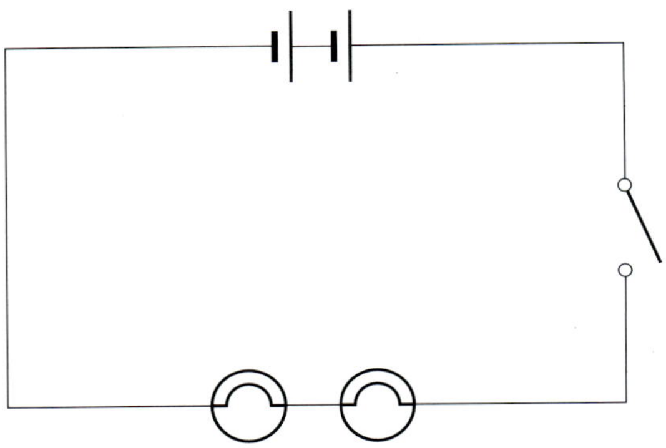

FIG 3.7.1 The symbols show that the circuit contains: two cells, two lamps and one switch connected in series

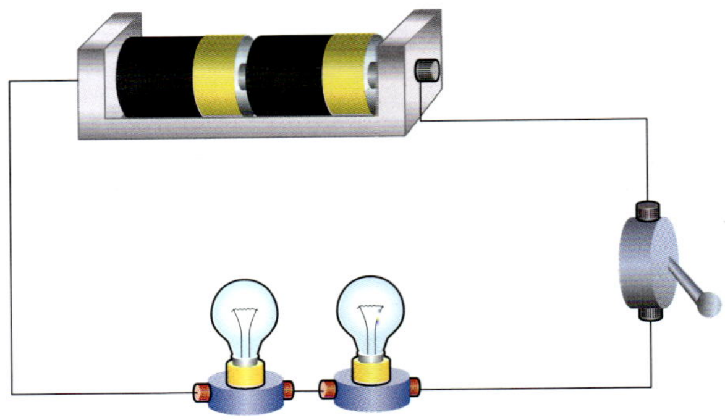

FIG 3.7.2 The information from the circuit diagram allows us to build the actual circuit

Activity 3.7.1

Building circuits from circuit diagrams

Here is what you need:

- three cells
- two lamps
- two switches
- ammeter
- connecting wires.

Here is what you should do:

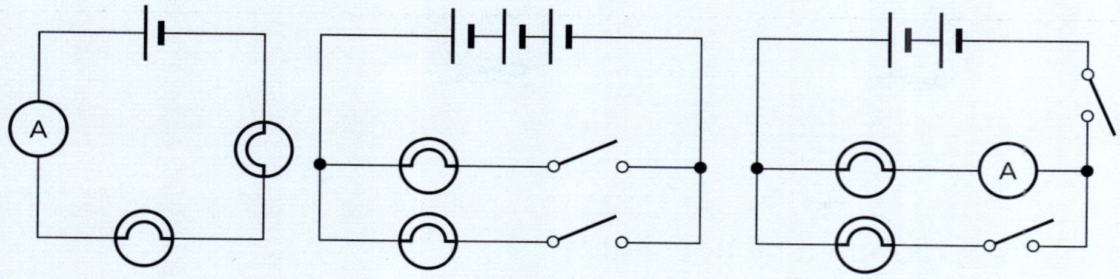

FIG 3.7.3

1. Select the components you will need to build circuit 1 and connect them together as shown in the diagram.
2. Check your circuit with the circuit diagram to make sure they are the same.
3. Repeat this for circuit 2 and then circuit 3.

Check your understanding

1. Fig 3.7.4 is a circuit diagram.

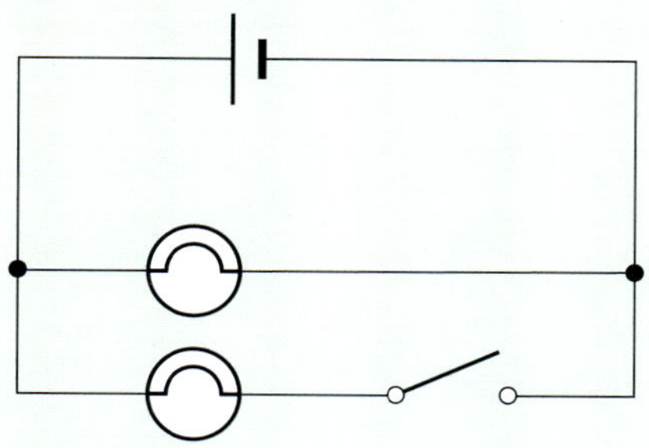

FIG 3.7.4

What components are needed to build this circuit (not including connecting wires)?

2. List the components you will need to build a circuit with three lamps in it, where each lamp can be switched on and off independently of the other two. Draw the circuit diagram and then build it.

Interesting fact

Sometimes wires may cross over each other in a circuit but may not actually join together.

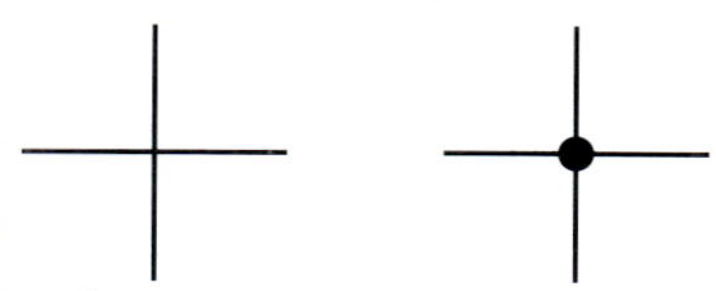

FIG 3.7.5 In order to show the difference between wires that cross and wires that join in a circuit diagram, we place a dot where wires join

3.8 Connecting components in series

We are learning how to:

- represent simple circuits using diagrams
- connect components in series in a circuit

The brightness of a lamp is determined by how much electrical energy is being converted to light (and heat) energy. The brightness is therefore a good indicator of the amount of current passing.

Activity 3.8.1

Investigating bulbs connected in series

Here is what you need:

- battery containing three cells
- four lamps
- ammeter
- connecting wires.

Here is what you should do:

1. Connect a single lamp in series with an ammeter in a circuit.
2. Note the brightness of the lamp and the reading on the ammeter.
3. Repeat this for two, three and four lamps connected in series and, in addition, find out what happens when one lamp is partially unscrewed from its holder so it goes out.
4. Record your observations in a table.
5. Comment on how the brightness of the lamps changes as the number increases.
6. Comment on how the current changes as the number of lamps increases.

FIG 3.8.1 What happens when increasing numbers of lamps are connected in series?

Lamps in series

As more lamps are added to the circuit, the lamps shine less brightly. If an ammeter is included in each circuit, it will show that the current falls as the number of lamps increases.

Lamps in **series** are connected by a single circuit. If there is a break in the circuit, such as will occur if one of the lamps burns out or is removed, then the circuit is no longer complete and all the lamps will go out.

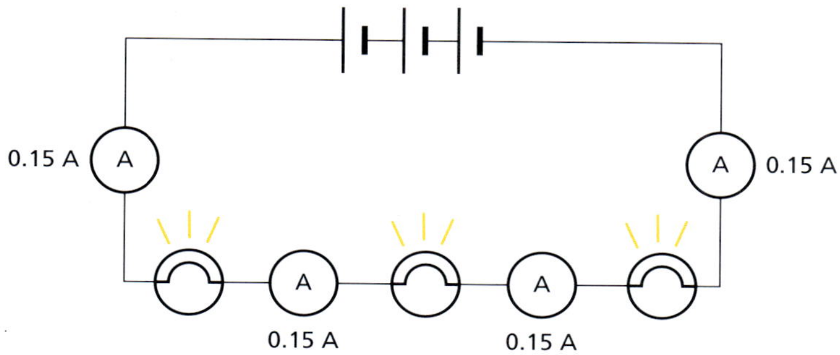

FIG 3.8.2 When lamps are connected in series, it does not matter where the ammeter is positioned as the current through the circuit is the same at all points (in this circuit the current is 0.15 A)

Check your understanding

1. Fig 3.8.3 shows a circuit in which three lamps are connected in series. The reading on the ammeter is 0.12 A.

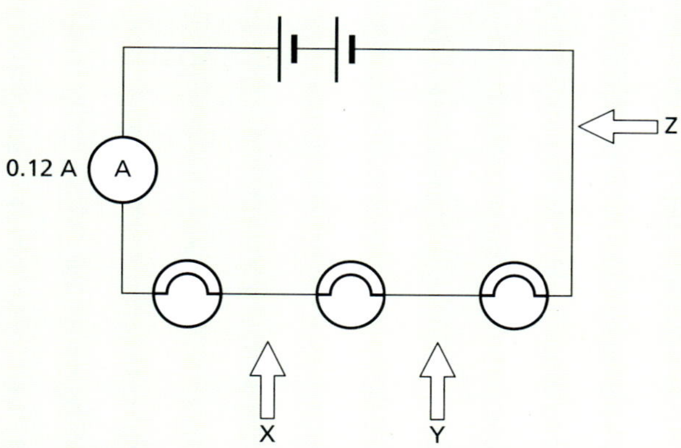

FIG 3.8.3

a) How would the brightness of the lamps change, if at all, if another lamp was added in series?

b) At which of the points X, Y and Z in the circuit would the current be 0.12 A?

c) What would be the reading on the ammeter if one of the lamps burned out? Explain your answer.

Interesting fact

When identical cells are connected in series in a battery, the potential difference across the battery is the sum of the potential differences of the cells, provided they are connected in the same direction.

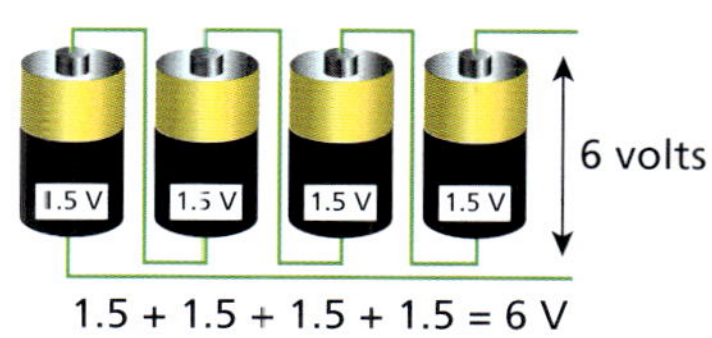

FIG 3.8.4

Key term

series way of connecting components so that they are in one loop in the circuit

3.9 Connecting components in parallel

We are learning how to:

- represent simple circuits using diagrams
- connect components in parallel in a circuit

An electric current only flows through a complete circuit. When two components are connected in **parallel**, there are effectively two circuits with a part that is common to both components.

Activity 3.9.1

Investigating lamps connected in parallel

Here is what you need:

- battery containing three cells
- four lamps
- connecting wires.

Here is what you should do:

1. Connect a single lamp in a circuit.
2. Note the brightness of the lamp.
3. Repeat this for two, three and four lamps connected in parallel and, in addition, find out what happens when one lamp is partially unscrewed from its holder so it goes out.
4. Comment on how the brightness of the lamps changes as the number increases.

Key term

parallel way of connecting components in a circuit so that the potential difference is the same across all branches of the circuit

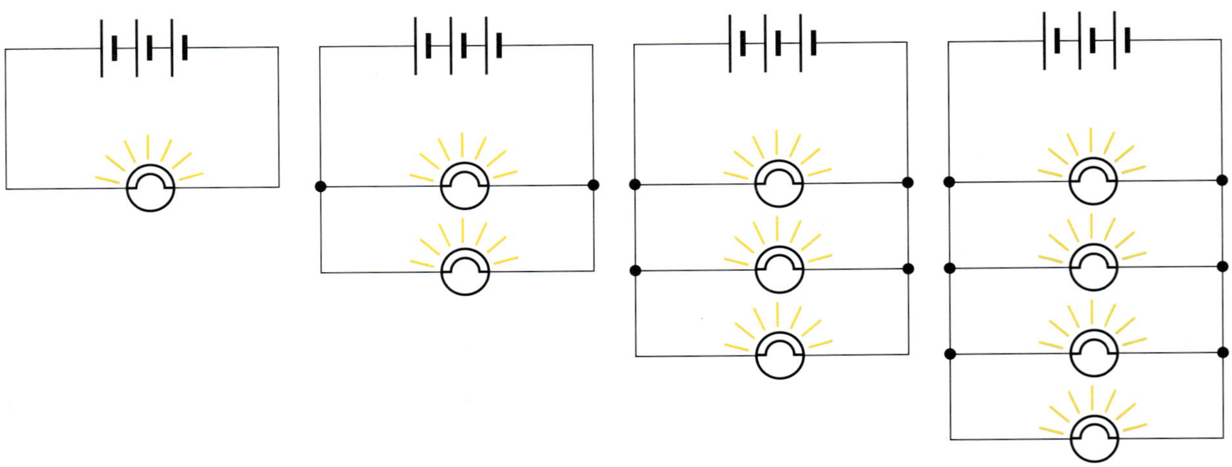

FIG 3.9.1 What happens when increasing numbers of lamps are connected in parallel?

Lamps in parallel

Adding lamps in parallel does not alter their brightness. However, it is not possible to keep on adding more and more lamps without end. There will come a time when the battery is unable to provide sufficient electrical energy.

When two lamps are connected in parallel, they are brighter than they would be if they were connected in series, but they draw twice as much current from the battery. This means that the battery will be exhausted more quickly. Two lamps connected in series will shine less brightly but will continue to shine for longer because the battery will last for longer.

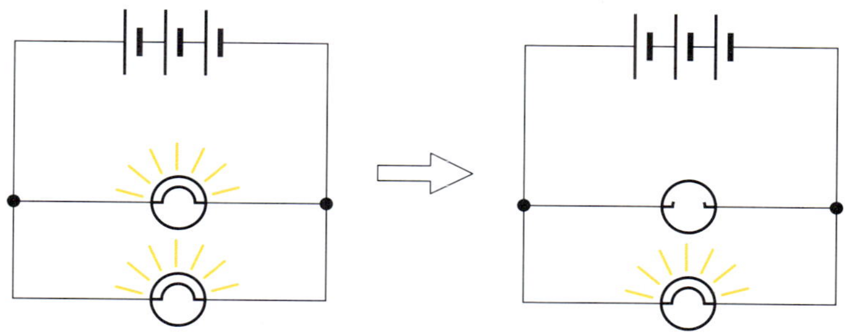

FIG 3.9.2 When lamps are connected in parallel, if one breaks, the circuit containing the second lamp remains complete and so the second lamp remains lit

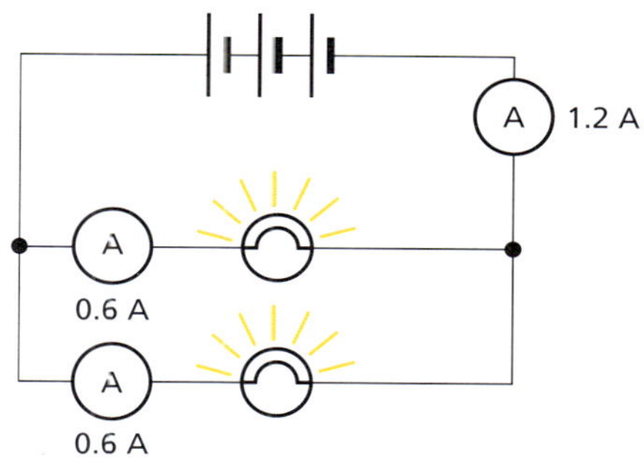

FIG 3.9.3 If ammeters are at different positions in the same circuit, the total current flowing from the battery is equal to the sum of the currents passing through each lamp, so here the values are: 0.6 A + 0.6 A = 1.2 A

Check your understanding

1.

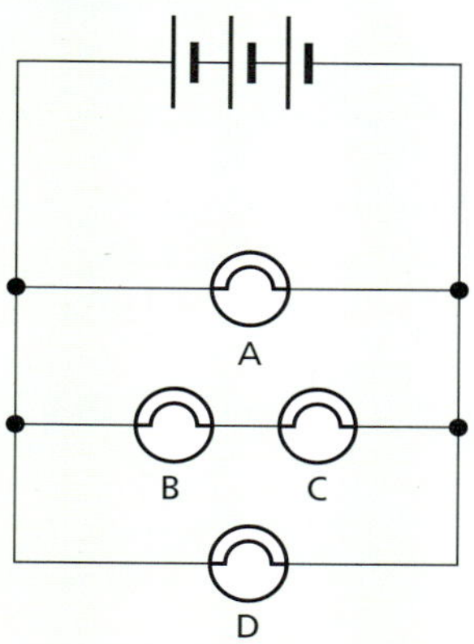

FIG 3.9.4 Four similar lamps connected to a battery

Copy and complete the table to show which lamps would remain on and which would go off when each of the lamps burned out.

Lamp that is burned out	Lamp goes out or remains on			
	A	B	C	D
A	off			
B		off		
C			off	
D				off

TABLE 3.9.1

Interesting fact

When identical cells are connected in parallel, the potential difference across them is the same as the potential difference across one cell provided they are connected in the same direction.

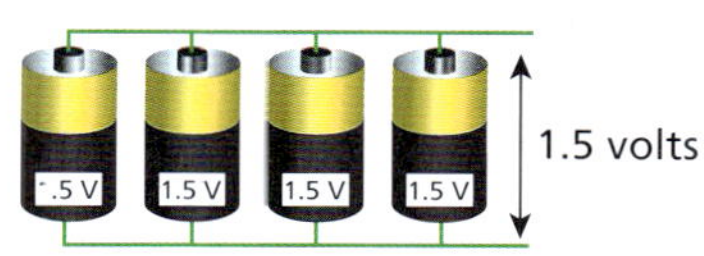

FIG 3.9.5

3.10 Resistance

All materials resist the flow of an electric current in a circuit. Materials which are good conductors, like metals, have a very low resistance, while materials which are insulators, like plastics and glass, have a very high resistance.

Components called resistors are used in electrical circuits to control the flow of current. They are conductors that have particular resistance values. The unit of resistance is the **ohm**, which has the symbol Ω.

There are two kinds or resistors:

- **fixed value resistors** – the value cannot be altered
- **variable resistors** – the value can be changed from zero to an upper limit.

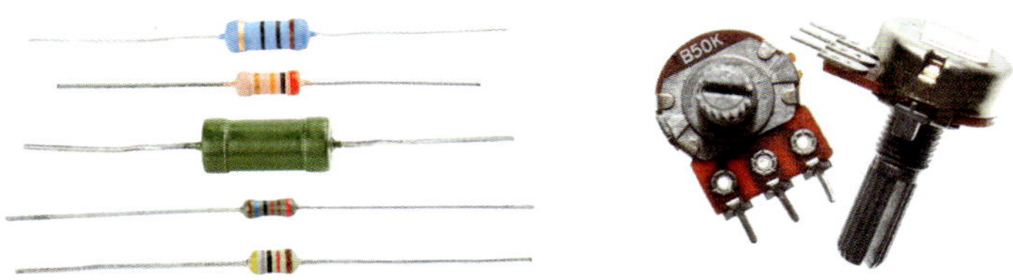

Fixed resistors

Variable resistor

FIG 3.10.1 Resistors

The values of fixed resistors are shown as a series of different coloured bands. You might have noticed them on the circuit board of an electrical device.

FIG 3.10.2 Symbol for a fixed resistor

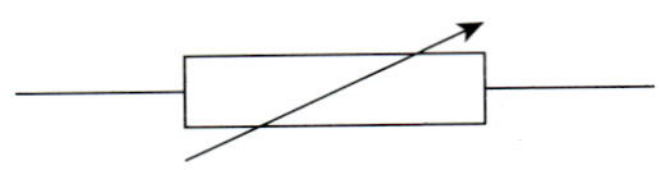

FIG 3.10.3 Symbol for a variable resistor

FIG 3.10.2 shows the symbol used to represent a fixed resistor in a circuit diagram. The value of the resistor is sometimes written in or under the component.

The value of a variable resistor can be changed within its given range. Variable resistors are sometimes called potentiometers and are commonly used as volume and other controls on devices such as radios and televisions.

Resistors provide a way of controlling the amount of current in a circuit, or in part of a circuit.

In the circuit in Figure 3.10.4, lamps A and B are identical. The current passing through lamp A is reduced by the fixed resistor. As a result, lamp A shines less brightly than lamp B.

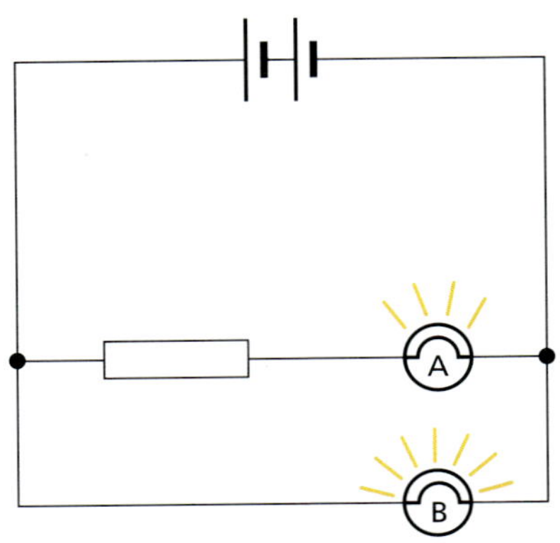

FIG 3.10.4 Resistors control current in a circuit

Activity 3.10.1

Effect of resistance on the brightness of a lamp

Here is what you will need:

- fixed resistors of different values – for example, 5, 10, 15, 20, 25, 30 Ω
- variable resistor – range 0–100 Ω
- battery of two 1.5 V cells
- lamp
- connecting wires.

Here is what you should do:

1. Connect a lamp to a battery consisting of two 1.5 V cells and note its brightness.
2. Connect the lowest value fixed resistor in series with the lamp and observe the effect this has on the brightness of the lamp.
3. Repeat step 2 using fixed resistors with progressively higher values.
4. Connect a variable resistor in series with the lamp and observe the effect that altering the value of the resistor has on the brightness of the lamp.
5. Use your observations to describe the effect of resistance on the brightness of a lamp in a circuit.

Interesting fact

The resistors used in practical circuits generally have values of thousands or even millions of ohms. As with all S.I. units, the prefixes 'k' and 'M' are used, i.e. kΩ = 1000 Ω and MΩ = 1 000 000 Ω.

Key terms

ohm unit of electrical resistance which has the symbol Ω

fixed resistor resistor whose value cannot be changed

variable resistor resistor whose value can be changed within given limits

Check your understanding

1. **a)** Draw a circuit diagram to show a fixed resistor and a lamp connected in series to a 1.5 V cell.

   b) Draw a circuit diagram to show a variable resistor and a lamp connected in series to a 1.5 V cell.
2. How does a resistor control the amount of current passing in a circuit?

3.11 Voltage and current in a circuit

We are learning how to:

- investigate the relationship between voltage (V) and current (I) in a simple series circuit

In an electric cell there is a **potential difference,** or a difference in energy levels between the two terminals. This potential difference is applied to any circuit to which the cell is connected. It is measured in volts (V) and for that reason is sometimes called the **voltage**.

Measuring voltage

Voltage is measured by a **voltmeter**. As was the case with ammeters, voltmeters may be analogue or digital.

A digital voltmeter may be part of a multimeter that can also be used as an ammeter depending on which function is selected.

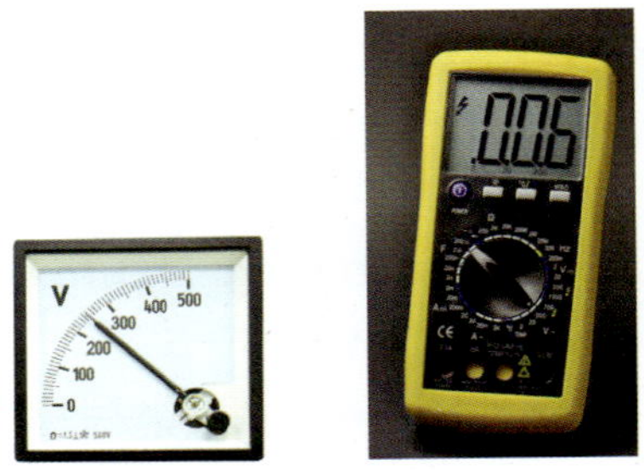

FIG 3.11.1 Voltmeters

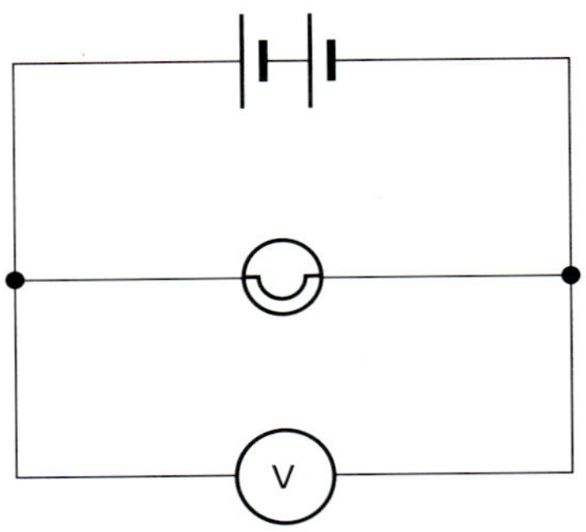

FIG 3.11.2 A voltmeter used in a circuit

A voltmeter is connected in parallel to a component to measure the voltage across it. When several components are connected in series in a circuit, the sum of the voltages across each component is equal to the supply voltage.

Current and voltage

Current and voltage are two different characteristics of the electrical energy passing around a circuit. An electric current only flows around a circuit if there is a voltage difference between the two terminals of the cell or other power source.

Interesting fact

The sizes of the current and voltage in a circuit are linked. If you change the voltage applied to a circuit, this will also alter the current that passes.

Activity 3.11.1

Investigating voltage and current in a circuit

Here is what you need:

- battery containing two cells
- lamp
- variable resistor
- ammeter
- voltmeter
- connecting wires.

A variable resistor is a component that resists the flow of current. Altering its value alters the size of its resistance.

Here is what you should do:

1. Connect the components to build the circuit shown in Fig 3.11.3.
2. Copy Table 3.11.1.

Voltage across the lamp	Current in the lamp	Brightness of the lamp

TABLE 3.11.1

3. Start by setting the value of the resistor to zero and record the voltage, current and brightness of the lamp.
4. Increase the value of the variable resistance to about one quarter, one half, three quarters and all of its value. Each time, record the voltage, current and brightness of the lamp.
5. Comment on your observations.

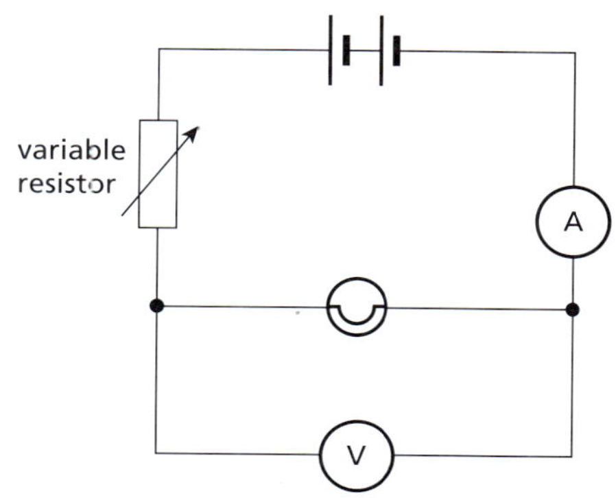

FIG 3.11.3

Check your understanding

1. Draw a circuit diagram in which there is:
   - a single lamp connected to a battery containing three cells
   - an ammeter to measure the current in the lamp
   - a voltmeter to measure the voltage across the lamp.
2. Four identical lamps were connected in series in a circuit as in Fig 3.11.4. Comment on the size of:

   a) the voltage across each individual lamp

   b) the total voltage across the lamps.

Key terms

potential difference the difference in energy levels at different points in an electrical circuit

voltage alternative name for potential difference

voltmeter device for measuring voltage

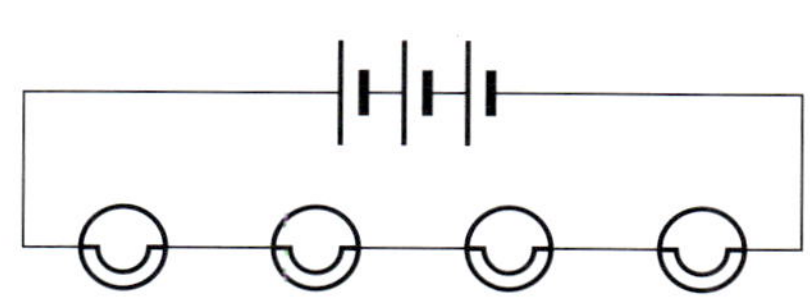

FIG 3.11.4

Review of Current electricity

- Electricity is a convenient form of energy.
- Materials that conduct an electric current are called electrical conductors.
- Materials that do not conduct an electric current are called electrical insulators.
- Metals are good conductors because they contain delocalised electrons. Their particles carry electrical charge along the metal. Insulating materials do not contain delocalised electrons.
- Electricity is potentially dangerous so electrical devices must be handled with care.
- Electrical appliances with metal bodies are connected to an earth wire. This provides a pathway for current in the event of wires coming loose and touching the metal body, which protects the user from a severe electric shock. Electrical appliances with plastic bodies do not require an earth wire and are said to be double insulated.
- A cell is a means of providing a small and safe amount of energy. A battery is formed when two or more cells are joined together.
- A cell has a positive (+) terminal and a negative (–) terminal. When cells are placed together to form a battery, they must all be pointing in the same direction.
- In order to flow, an electric current must have a complete circuit.
- An electric current can be measured using an ammeter. The unit of current is the ampere, A, or for smaller currents the milliampere, mA. There are 1000 milliamperes in 1 ampere.
- A circuit diagram is a method of representing a circuit by a series of connected symbols. Each symbol represents a component in the circuit. Circuit diagrams are quick to draw and will be universally understood because scientists around the world use the same symbols.
- Components in a circuit might be connected in series or in parallel.
- When lamps are connected in series, the more lamps there are the dimmer they are and the less current flows in the circuit.
- When lamps are connected in parallel, they shine with the brightness of a single lamp up to the point where the cell or battery cannot provide any more electrical energy.
- Lamps connected in parallel are brighter than lamps connected in series but they draw more current from the battery. This means that the battery will be exhausted more quickly. Lamps connected in series will shine less brightly but will continue to shine for longer because the battery will last longer.
- Resistance is measured in ohms and has the symbol Ω.
- Resistors can be used to control the flow of current in a circuit.
- There is a difference in potential energy or voltage between the terminals of a cell and this is expressed in volts.
- Voltage is measured in volts using a voltmeter.
- When the voltage across a circuit changes the current in it will also change.

Review questions on Current electricity

1. Redraw the circuit in Fig 3.R.1 as a circuit diagram, using appropriate symbols for the components.

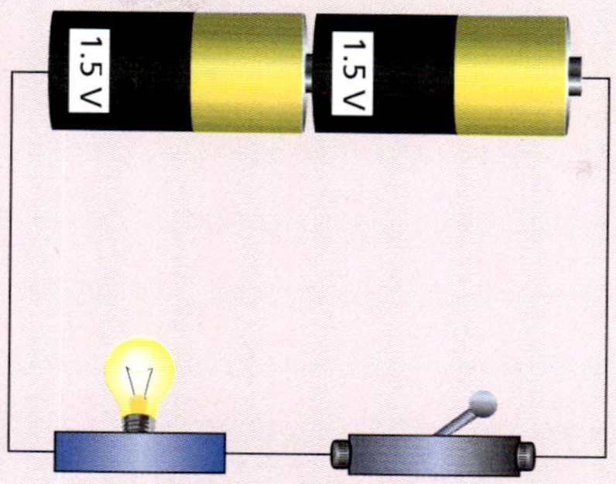

FIG 3.R.1

2. Dante used the circuit shown in Fig 3.R.2 to test whether materials conduct electricity.

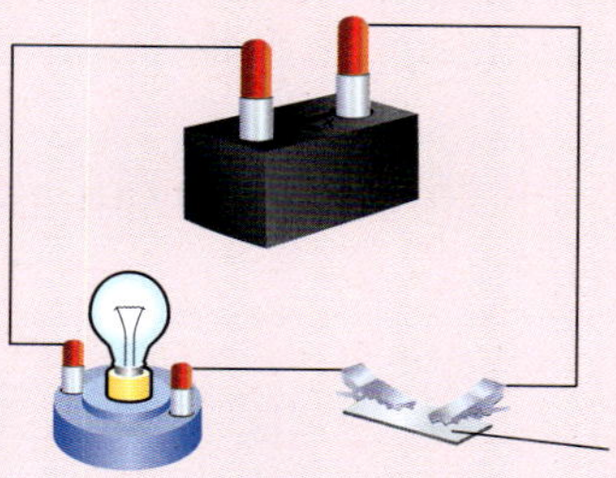

FIG 3.R.2

a) How was Dante able to decide whether or not each material conducted electricity?

b) Before testing the materials, Dante connected the two crocodile clips together. Why did he do this?

c) Here is a list of the materials Dante tested. Arrange them in groups according to whether or not they conduct electricity.

copper	cardboard	plastic	iron
wood	glass	steel	lead

d) What name is given to a material that does not conduct electricity?

e) Give one common feature of all of the materials that conduct electricity.

3. Fig 3.R.3 shows an experiment using some circuits containing identical lamps.

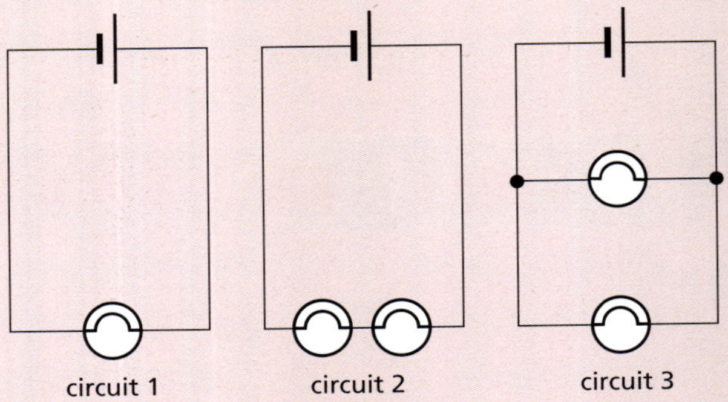

FIG 3.R.3

a) If the brightness of the lamp in circuit 1 is 'normal', how bright are the lamps in circuit 2 and circuit 3 compared to circuit 1?

b) If the cells used in these circuits are identical at the start of the experiment, how long will the cell in circuit 2 and in circuit 3 last compared to the cell in circuit 1?

c) Explain what will happen if one of the lamps burns out in:

i) circuit 2 **ii)** circuit 3.

4. Fareed wants to make a circuit containing two cells, a lamp and a switch so that the lamp lights up when the switch is closed, but he is having problems.

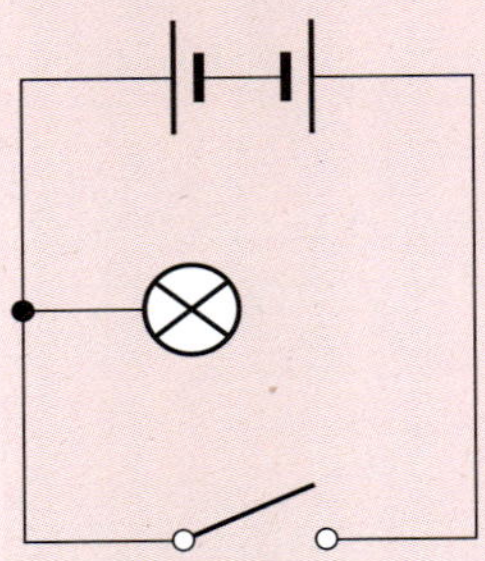

FIG 3.R.4

a) There are three problems with Fareed's circuit. Explain what they are.

b) Redraw the circuit showing the components drawn correctly.

5. Fig 3.R.5 shows a cell and two lamps connected in different ways.

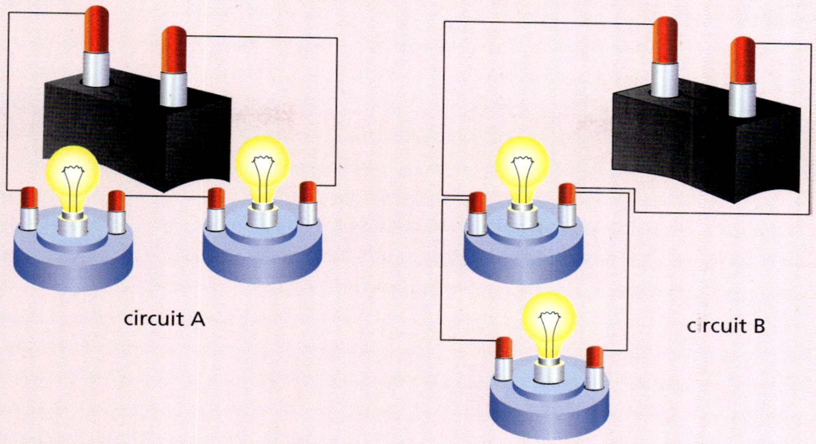

FIG 3.R.5

a) Redraw the circuits as circuit diagrams.

b) Compared to the brightness of a single lamp connected to a cell, describe the brightness of:

i) the lamps in circuit A

ii) the lamps in circuit B.

c) What would happen to the second lamp if one lamp were to break in:

i) circuit A?

ii) circuit B?

6. In the following circuit, lamps P, Q and R are identical and components X are identical.

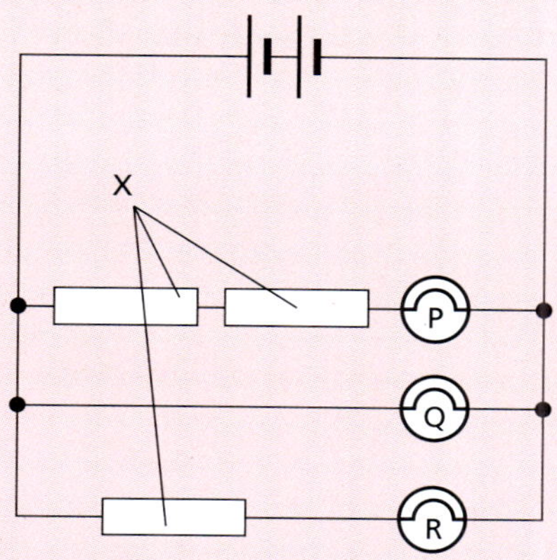

FIG 3.R.6

a) What is the name of component X?

b) Write down the order in which the lamps shine starting with the brightest.

Science in practice

Lighting system for a model house

Before you start work on this activity, you should recall that mains electricity is dangerous. You should not remove the tops from lamps and switches or expose wires in mains lighting circuits.

Mr Livingston is making a model doll house as a surprise for his daughter's birthday.

FIG SIP 3.1 The model doll house Mr Livingston is making

He wants to make it extra special by having a light in the ceiling of each room and a switch by the door that can turn it on and off, but he doesn't know anything about electricity.

1. You are going to work in a group of three or four to design a lighting system for Mr Livingston's doll house, to make the components and to make a model to show how your system works. The tasks are:

- to research about house lighting
- to design a suitable lighting circuit
- to design and make lamps and switches to a suitable scale for a doll house
- to build a model of a doll house out of cardboard
- to install your lighting system in your model
- to test your lighting system
- to consider how your lighting system might be modified or extended.

a) Take a look at how the lighting is arranged in your home.

In each room the switch only controls the light for that room.

Use your knowledge of series and parallel circuits to design a suitable lighting circuit for a doll house that contains four rooms.

FIG SIP 3.2 Room light and switch

b) Even small torch bulbs are too big for a doll house. You need to find lights which are smaller than these.

One possibility is light-emitting diodes. These are not bulbs but components that shine when a current passes through them. Red LEDs are often used on electrical equipment to show that something is switched on. They come in three sizes – 3 mm, 5 mm and 10 mm – so you will need to choose which is appropriate for the scale of your model.

If you use LEDs there are two important things you must take into account:

- LEDs cannot be connected directly to cells or they will burn out instantly. They must be wired in series with a resistor. You will need to carry out research to determine the size of the resistor.
- LEDs must be connected in a circuit so that the longer pin is connected to the positive side of the cell or battery. Otherwise they will not work.

Similarly, household switches are far too large for a doll house so you will have to design and make your own.

The examples of homemade switches in Fig 3.4 might give you some ideas. You will need to make a switch for each room of the model doll house.

c) Build a model of a doll house using a cardboard box and cardboard dividers for the walls.

Your model doesn't have to be very detailed as you are only going to use it to demonstrate your lighting system.

d) Install your lighting system into your model and test it to make sure the light in each room turns on and off without affecting the lights in the other rooms. Use a battery to power your lights.

e) Consider how you might modify or extend your lighting system. For example:

- You might install two switches in one room so that the light could be controlled from either switch. This is the sort of arrangement that you see in hallways and staircases.
- You might install two lights in one room controlled by only one switch.

f) Prepare a presentation in which you will describe your lighting circuit, explain how you made lights and switches, and demonstrate the lighting circuit installed on your model. You should also be ready to discuss how your lighting circuit could be modified to provide different arrangements of lights and switches.

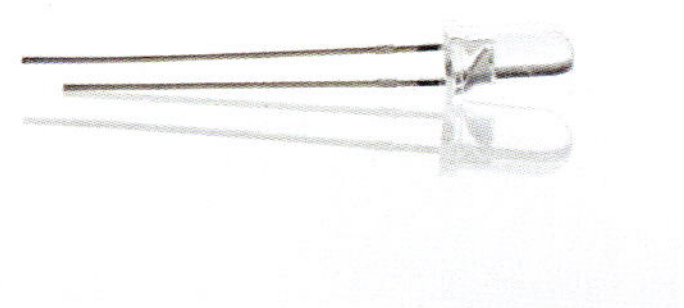

FIG SIP 3.3 Light-emitting diode or LED

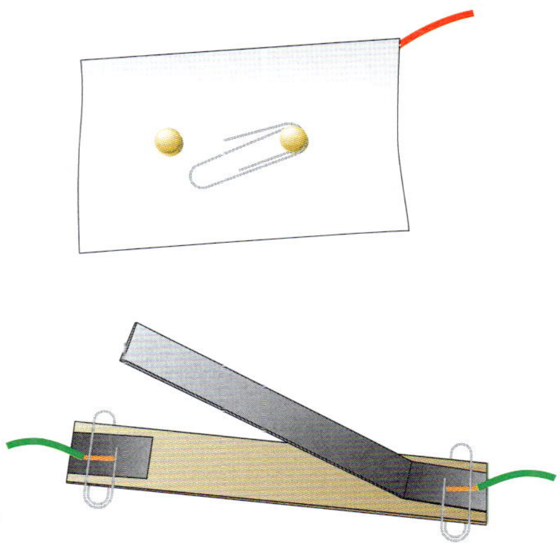

FIG SIP 3.4 'Homemade' switches

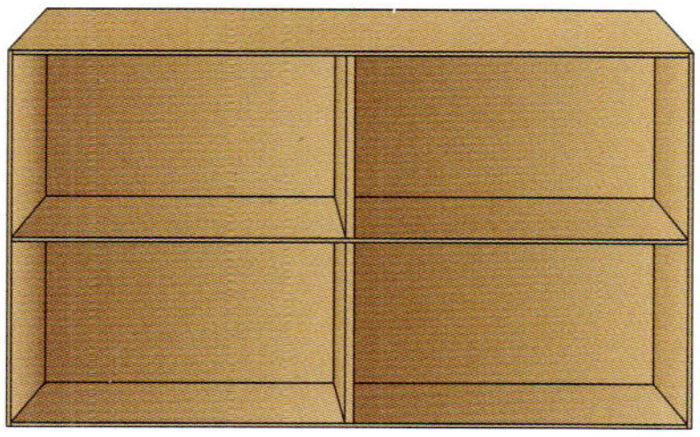

FIG SIP 3.5 Model doll house

Unit 4: Magnetism

4.1 Magnetic and non-magnetic materials

We are learning how to:

- demonstrate the effects of magnetic forces
- determine whether a material is magnetic or not.

Lodestone is an oxide of iron and is also called magnetite.

FIG 4.1.1 Lodestone, a type of rock, attracts objects made of iron, such as nails and bolts

Activity 4.1.1

Magnetic and non-magnetic materials

Here is what you need:

- magnet
- objects made of different materials – for example, nail, paper clip, plastic ruler, eraser.

Here is what you should do:

1. Place one end of the magnet near an object and find out if the material is attracted to it.
2. Materials that are attracted by a magnet are described as magnetic materials.
3. Test each object in turn with the magnet.

4. Display your observations in a table. On one side of the table list the magnetic materials and on the other side list the non-magnetic materials.

Materials such as iron and steel that are attracted to a magnet are described as **magnetic**, while materials such as brass, copper and aluminium that are not attracted to a magnet are described as **non-magnetic**.

FIG 4.1.2 Ceramic magnets can be made in any shape but have the disadvantage that they are brittle, so if they are dropped on a hard surface they will break into pieces

Permanent magnets

Materials that keep their magnetism for a long time are called permanent magnets.

Materials that have permanent magnetism are iron, mild steel, cobalt and nickel. Modern magnets are often made of special alloys containing these metals, such as alnico and alcomax.

Ceramic or ferrite magnets are made by baking iron oxide and other metal oxides in a ceramic matrix.

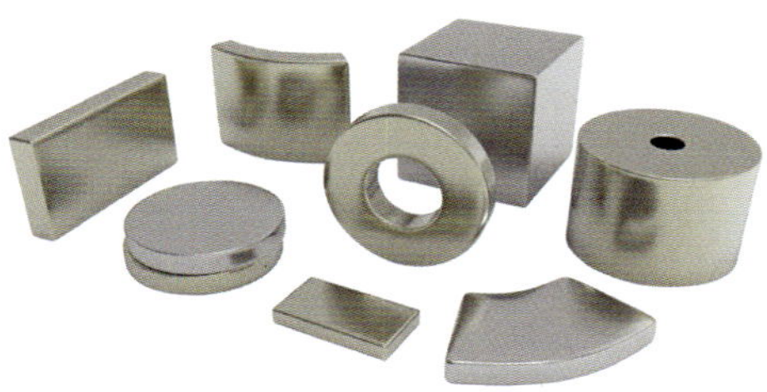

FIG 4.1.3 The element neodymium forms alloys with iron and boron that are used to make powerful permanent magnets

Interesting fact

An alloy is a mixture of a metallic element with one or more other elements that may be metals or non-metals. Steel is an alloy of iron and carbon.

Sometimes alloys have more useful properties than the elements from which they are formed. For example, alnico is an alloy of aluminium, nickel and cobalt. It makes more powerful magnets than the pure metals.

Check your understanding

1. Arrange the following metals into two groups: those that are magnetic and those that are not magnetic.

cobalt	copper	gold	iron
magnesium	nickel	steel	zinc

Key terms

magnetic materials such as iron and steel that are attracted to a magnet

non-magnetic materials such as brass, copper and aluminium that are not attracted to a magnet

4.2 Law of magnetic poles

We are learning how to:

- demonstrate the effects of magnetic forces
- predict whether two magnetic poles will attract or repel each other.

Poles on a magnet

A magnet has two **poles**: a north pole and a south pole. The north and south poles are usually represented by the letters 'N' and 'S'.

Forces exist between magnets and are concentrated at the poles. The interaction between two magnets depends on the nature of the poles that are brought together.

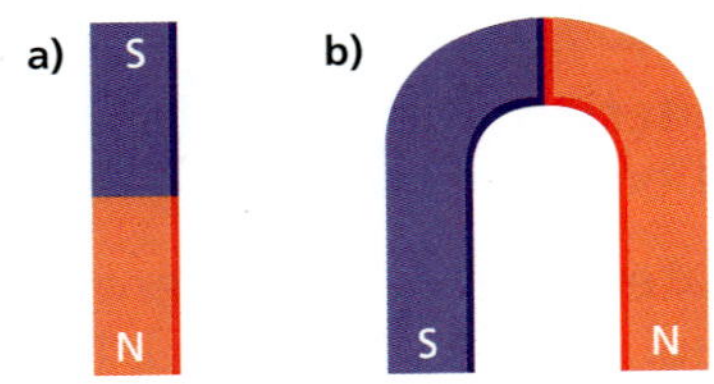

FIG 4.2.1 **a)** Bar magnets are commonly used in the laboratory **b)** A horseshoe magnet is simply a bar magnet that has been bent into the shape of a horseshoe

Activity 4.2.1

Law of magnetic poles

Here is what you need:

- two loops of cotton
- two bar magnets
- pencil
- heavy book.

Here is what you should do:

1. Place a heavy book on top of a pencil so that the pencil is sticking out from the table.
2. Suspend a bar magnet from the pencil using loops of cotton so that it can turn freely.

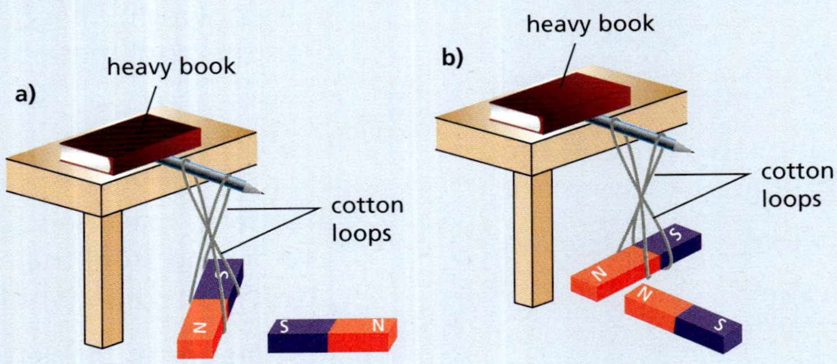

FIG 4.2.2 **a)** Unlike poles **b)** Like poles

3. Bring the N pole of the second magnet towards the N pole of the suspended magnet and record what happens.
4. Bring the N pole of the second magnet towards the S pole of the suspended magnet and record what happens.

5. Repeat steps 3 and 4 but using the S pole of the second magnet.
6. What deductions are you able to make about magnets from your observations?

If one magnet is suspended so it is free to rotate and a second magnet is brought near it:

- If they are **unlike poles**, that is N and S or S and N, the magnets will attract (move towards each other).
- If they are **like poles**, that is N and N or S and S, the magnets will repel (move away from each other).

Interesting fact

It is impossible to say if an iron bar is magnetic or not on the basis of whether it is attracted by a magnet.

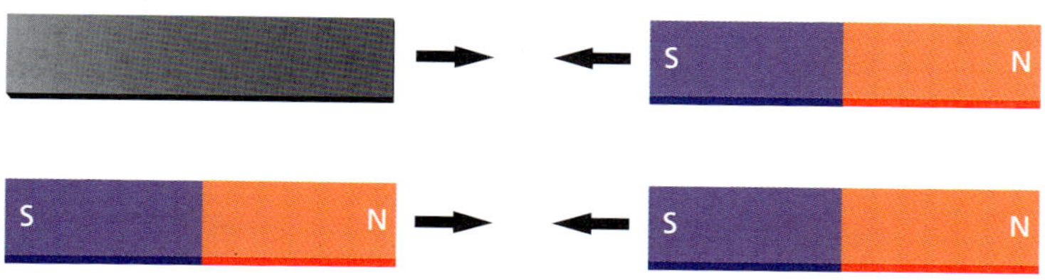

FIG 4.2.3 A magnet would attract an iron bar even if the iron bar was not itself a magnet

FIG 4.2.4 Repulsion proves an iron bar is a magnet

To test whether an iron bar is a magnet, both ends must be placed near the same magnetic pole of a magnet. If one end is repelled, this proves that the bar is a magnet.

Key terms

pole the end of a magnet

unlike poles two poles that are different, that is, north and south

like poles two poles that are the same, that is, north and north

Check your understanding

1. A compass needle is a magnet. The north pole of the compass always points towards the Earth's magnetic North Pole, and the south pole of the compass points towards the Earth's magnetic South Pole.

   What is the polarity of each of the Earth's magnetic poles? Explain your answer.

FIG 4.2.5 Compass

4.3 Magnetic fields

We are learning how to:

- demonstrate the effects of magnetic forces
- draw the magnetic field around a bar magnet.

A magnet is surrounded by a pattern of invisible **magnetic field lines**. We can investigate the nature of the field lines using a small plotting compass.

Activity 4.3.1

Magnetic field around a bar magnet

Here is what you need:

- bar magnet
- plotting compass
- plain paper.

Here is what you should do:

1. Place a bar magnet at the centre of a sheet of plain paper and draw around it.
2. Remove the magnet and mark the N and S poles on the outline.
3. Place the magnet back on the outline.
4. Place the plotting compass near the north pole of the magnet. Mark two dots on the paper corresponding to the ends of the plotting compass needle.
5. Move the compass a little bit away towards the south pole and repeat drawing the dots.
6. Repeat this procedure until a circle is formed around the magnet.

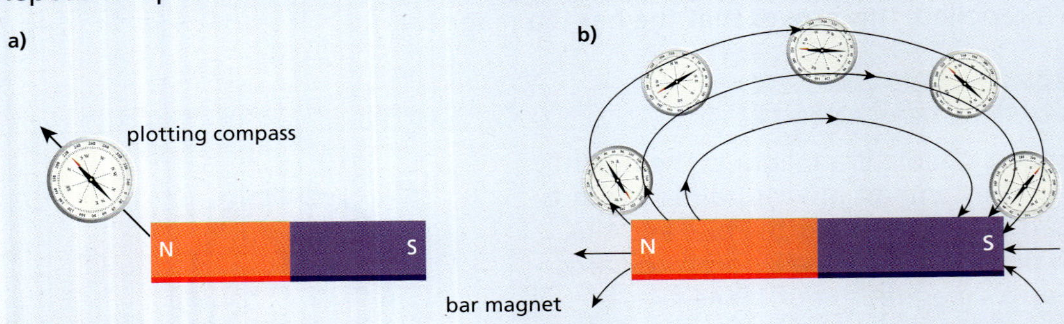

FIG 4.3.1

7. Connect all the dots with a smooth curve. This curve is one magnetic field line. Try to obtain four curves on each side of the magnet.

Drawing magnetic fields

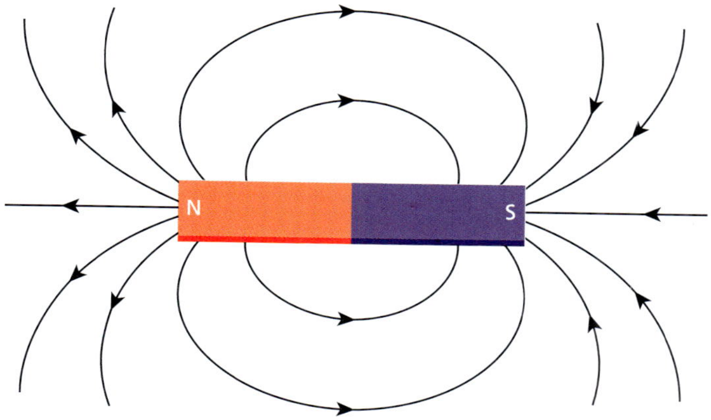

FIG 4.3.2 The magnetic field around a magnet can be represented by a set of magnetic field lines

Magnetic field lines are conventionally shown moving away from a north pole and towards a south pole, and arrows are placed on the lines to show the direction of the field. When drawing or interpreting the magnetic field around a magnet you should remember that:

- Magnetic field lines never cross over each other.
- The **magnetic field strength** is shown by the concentration of field lines and is strongest where the field lines are most dense (at the poles).
- The magnetic force of a magnet decreases with distance from the poles.

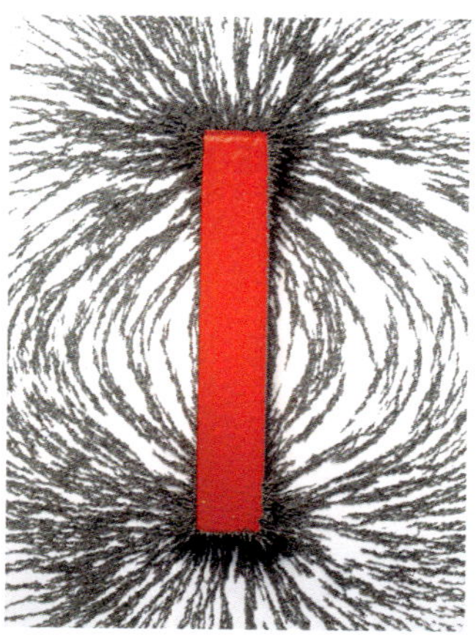

FIG 4.3.4 The magnetic field lines around this bar magnet are also shown by the iron filings around the magnet

Interesting fact

When two like poles are brought together, the field lines from each magnet run away from each other. At the centre of the combined magnetic fields is a null point at which there is no net magnetic field.

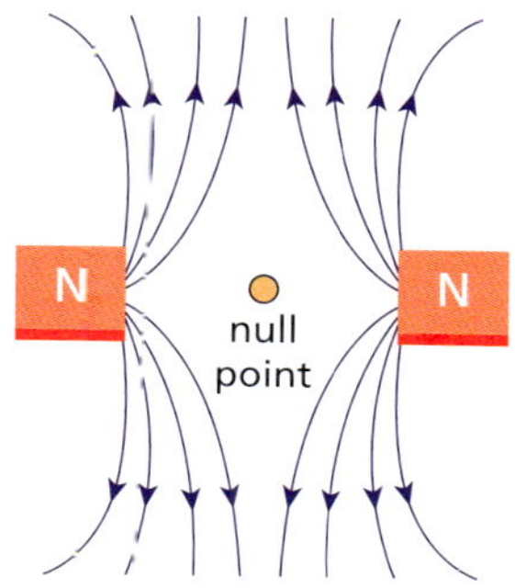

FIG 4.3.3 Magnetic field lines for unlike poles

The pattern of magnetic field lines for two S poles is the same as Fig 4.3.3, but the field lines run in the opposite direction.

Key terms

magnetic field lines a pattern of invisible lines that shows how the strength and direction of the magnetic field varies around a magnet

magnetic field strength how strong the magnetic field is at a particular point

Check your understanding

1. Use what you have learned about the magnetic field lines around a bar magnet to draw the magnetic field between the two poles of a horseshoe magnet.

4.4 Magnetic effect of an electric current

We are learning how to:

- describe the magnetic effect of current
- make an electromagnet.

When a compass needle is placed close to a wire, and then a current is passed through the wire, the compass needle is deflected.

This is called the **magnetic effect of a current**.

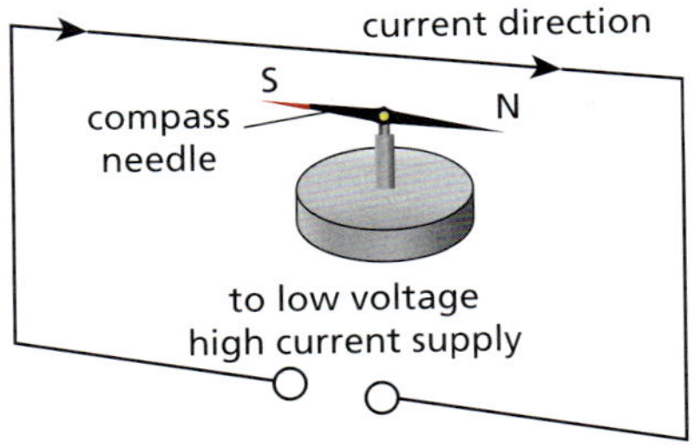

FIG 4.4.1 The magnetic effect of an electric current

Activity 4.4.1

Magnetic field around a wire carrying a current

Here is what you need:

- thick resistance wire
- plotting compass
- plain card
- DC power source.

Here is what you should do:

1. Make a small hole in the middle of a piece of card and push the wire through it.
2. Connect the wire to a DC power source. This has a positive (+) terminal and a negative (–) terminal.

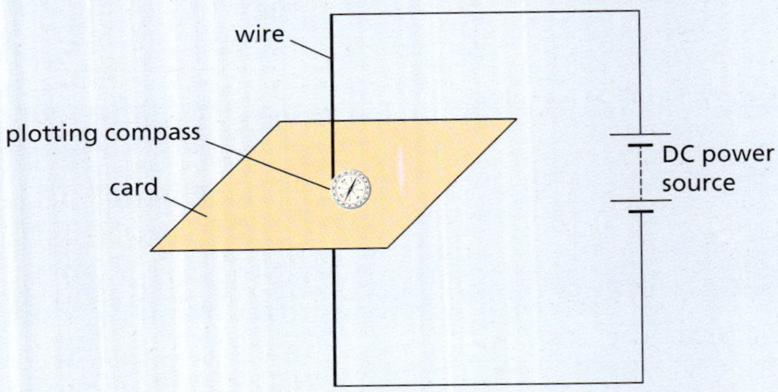

FIG 4.4.2

3. Place a plotting compass near the wire and show the direction that the compass points in by drawing an arrow.
4. Repeat this, placing the plotting compass in different positions until you have built up a map of the field lines around the wire.
5. Reverse the direction of the current through the wire by connecting the wire to the opposite terminals of the power source.
6. Observe if this affects the shape of the magnetic field lines around the wire and the direction of the magnetic field.

Strength of magnetic field

Passing a current through a conductor such as a wire creates a magnetic field consisting of a series of concentric circles. The circles are closer together nearer the wire where the magnetic field is strongest.

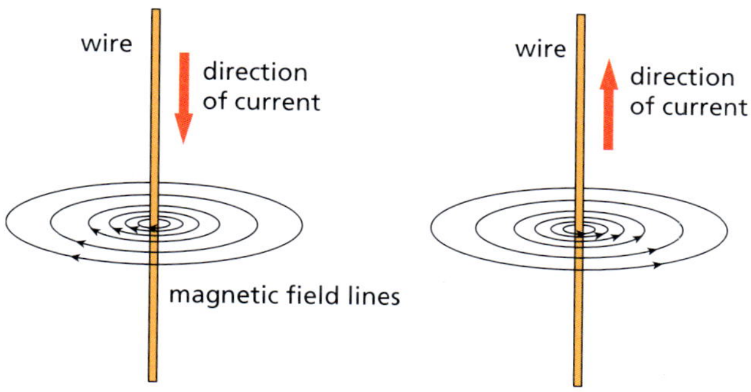

FIG 4.4.3 Magnetic field around a wire carrying an electric current

Reversing the direction of the current in the wire does not alter the shape of the magnetic field, but it does alter the direction of the field lines.

Check your understanding

1. Draw a diagram showing the magnetic field around a wire carrying a current, viewed as if you were looking along the wire.

Interesting fact

A solenoid is a coil consisting of a number of loops of wire.

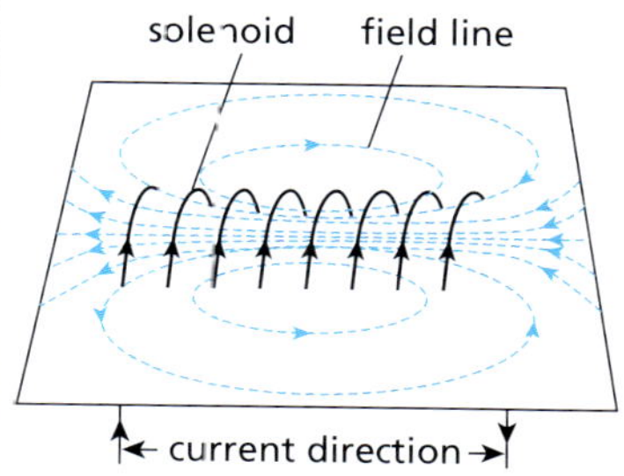

FIG 4.4.4 Magnetic field lines around a solenoid

The magnetic field around the solenoid is similar in shape to the magnetic field around a bar magnet.

Key term

magnetic effect of a current a compass needle is deflected when it is placed close to a wire carrying an electric current

4.5 Making an electromagnet

We are learning how to:

- describe the magnetic effect of current
- make an electromagnet

To make an **electromagnet** of any useful strength, you need to combine the magnetic field around many turns of wire by making a coil or solenoid.

The coils of wire on their own are magnetic. However, if they are wrapped around a steel nail, they make an even stronger magnet.

Activity 4.5.1

Making an electromagnet

Here is what you need:

- steel nail
- plastic-coated wire
- DC power source
- paper clips
- plotting compass.

Here is what you should do:

1. Take a length of wire that is coated in plastic insulation and coil it round a steel nail.
2. Make between 15 and 20 coils of wire round the nail, depending on the length of the wire.

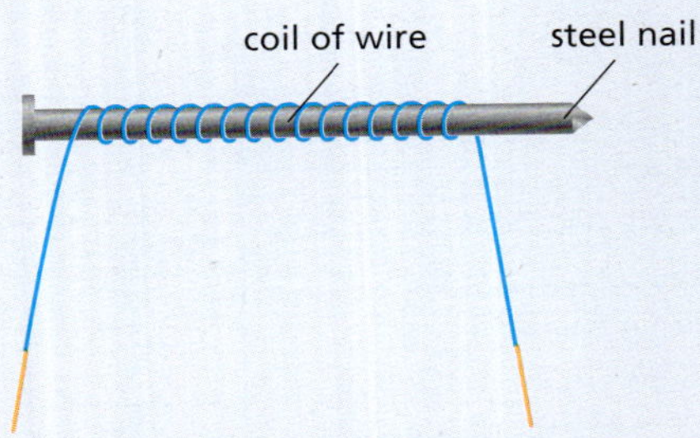

FIG 4.5.1 Coils of wire around a nail

3. Connect your coil to a DC power source.
4. Check that you have made an electromagnet by seeing if metal paper clips are attracted to it.
5. Place a plotting compass at different points around your electromagnet and use the direction the compass points to each time to draw a diagram of the magnetic field lines around your electromagnet.

Temporary magnetism

Electromagnetism is sometimes described as temporary magnetism. An electromagnet is only magnetic while a current flows through it. If the current is turned off, the electromagnet ceases to be magnetic.

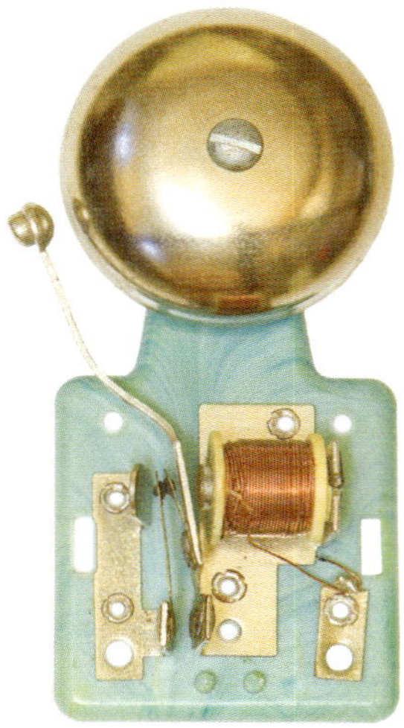

FIG 4.5.2 A practical electromagnet

Electromagnets used in devices such as electric bells consist of many coils of thin copper wire. At first glance the wire might not appear to be insulated, but it is. The wire is covered in a layer of lacquer, which is far less bulky than a plastic coating.

Check your understanding

1. Explain why a coil of wire can only attract paper clips when an electric current is passing through it.

Interesting fact

William Sturgeon made the first electromagnet in 1824.

FIG 4.5.3 William Sturgeon's first electromagnet consisted of about 18 turns of varnished wire wrapped around a piece of iron in the shape of a horseshoe

Key term

electromagnet a magnet produced when a current is passed through a wire or coil of wire

4.6 Strength of an electromagnet

We are learning how to:

- describe the magnetic effect of current
- compare the strengths of different electromagnets

An electromagnet is a coil of wire through which an electric current is passed. What determines the strength of an electromagnet?

- Would wrapping the coil round a piece of wooden dowel be just as good as wrapping it round a steel nail?
- Does it matter how many turns of wire are in the coil?
- Does it matter how much current you pass through the coil?

Activity 4.6.1

Investigating the strength of electromagnets

Here is what you need:

- steel nail
- wooden dowel
- plastic-coated wire
- DC power source
- paper clips.

Here is what you should do:

1. Take a length of wire that is coated in plastic insulation and coil it round a steel nail.
2. Make 20 coils of wire round the nail.
3. Connect your electromagnet to a DC power supply and count how many paper clips it will lift off the desk.

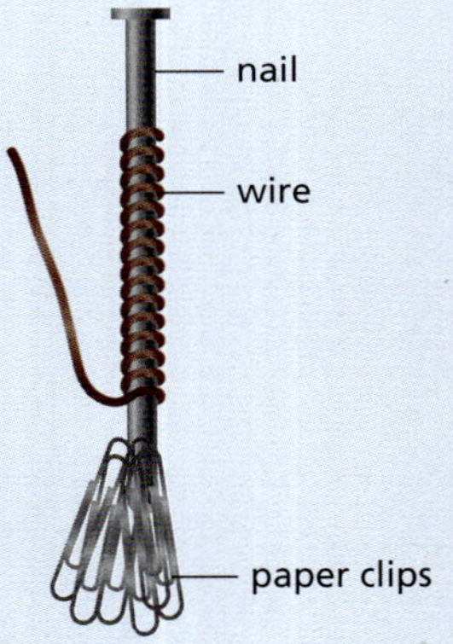

FIG 4.6.1

4. Repeat steps 1 to 3 but use a wooden dowel in place of a steel nail.
5. Now make 10 coils of wire round a steel nail.
6. Connect your electromagnet to a DC power supply and count how many paper clips it will lift off the desk.
7. Repeat steps 2 to 6, but this time use only half of the current used previously.
8. From your observations, deduce what factors determine the strength of an electromagnet.

Increasing the strength of an electromagnet

The strength of an electromagnet is increased by:

- wrapping the coils of wire round a core of a magnetic metal such as iron or steel
- increasing the number of turns of wire in the coil
- increasing the amount of current passing through the coil.

Check your understanding

1. Say whether each of the following statements is true or false.

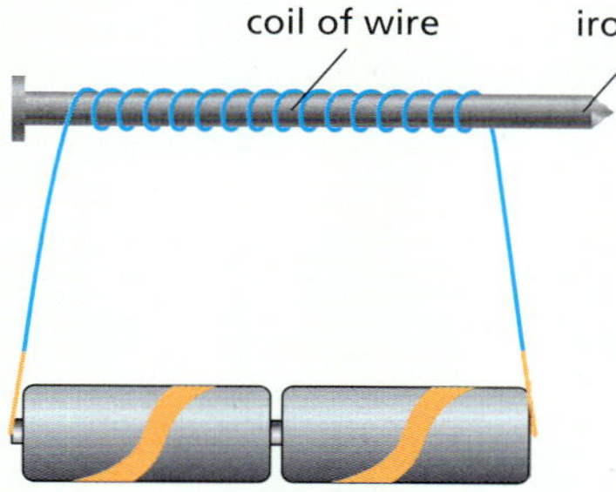

FIG 4.6.2 An electromagnet made by wrapping wire around an iron nail and connecting it to a battery

   a) Reversing the battery will reduce the strength of the electromagnet.
   b) The iron nail would still be a strong magnet even if the battery was removed.
   c) Decreasing the number of turns of wire on the coil would reduce the strength of the electromagnet.
   d) The electromagnet would be stronger if the iron nail was removed.
   e) Wrapping the wire around two iron nails would make the electromagnet twice as strong.
   f) Connecting the coil to a battery with a higher voltage would make the electromagnet stronger.

Interesting fact

Credit and debit cards have a magnetic strip containing information about the cardholder's account.

FIG 4.6.3 If a card with a magnetic strip is placed too near to another magnet, the magnetic field will corrupt the information and the card will no longer be of any use

4.7 Uses of permanent magnets and electromagnets

We are learning how to:

- describe the magnetic effect of current
- explain the uses of permanent magnets and electromagnets

Permanent magnets and electromagnets have many applications, including the electric bell, the **relay** switch and the **circuit breaker**.

Electric bell

Activity 4.7.1

Investigating how an electric bell works

Here is what you need:

- electric bell
- power source
- screwdriver.

Here is what you should do:

1. Remove the cover from the outside of the bell so you can examine the parts.
2. Identify the electromagnet.
3. Turn the bell on and off and observe the effect this has on the components.
4. Use your knowledge of electric circuits and magnets to explain how the bell works.

When the bell switch is pushed, the circuit is complete and the following happens:

- The electromagnet become magnetic.
- The electromagnet attracts the soft iron **armature** and the hammer strikes the gong.
- As the soft iron armature moves, the circuit is broken and the electromagnet loses its magnetism.
- The springy metal strip moves the armature back to its starting position and the cycle repeats for as long as the switch is pushed.

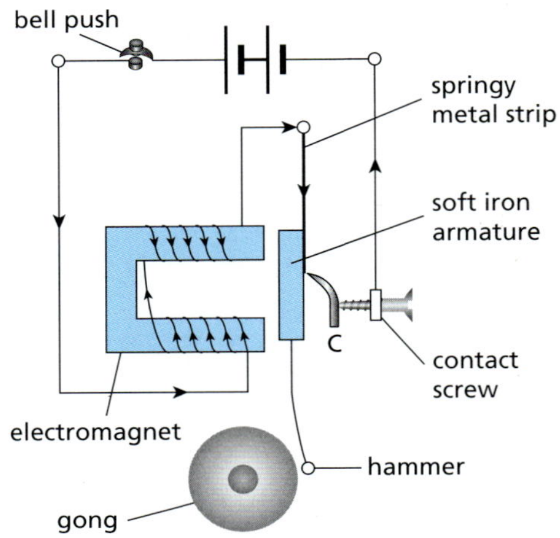

FIG 4.7.1 The structure of an electric doorbell

Relay

A relay is a switch that is operated by an electromagnet.

When a current passes between A and B, the soft iron core becomes an electromagnet. The iron armature is attracted to the electromagnet and turns on the pivot. The springy metal contacts at C are closed, completing the circuit connected across D and E.

A relay allows a circuit carrying a large current to be controlled by a second circuit carrying a small current. For example, a large current is needed to start a car engine. It is activated by the ignition switch in the car, which only carries a small current. This means that only thin wires are needed for the ignition circuit.

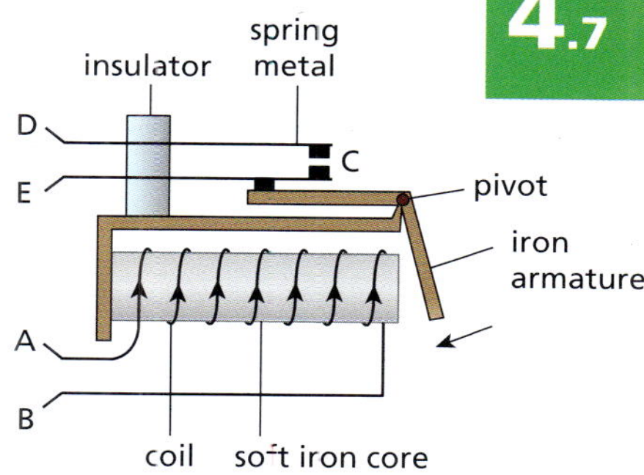

FIG 4.7.2 A relay allows one circuit to be controlled by another

Circuit breaker

A circuit breaker is a device that prevents the flow of current in a circuit in the event of a malfunction or fault.

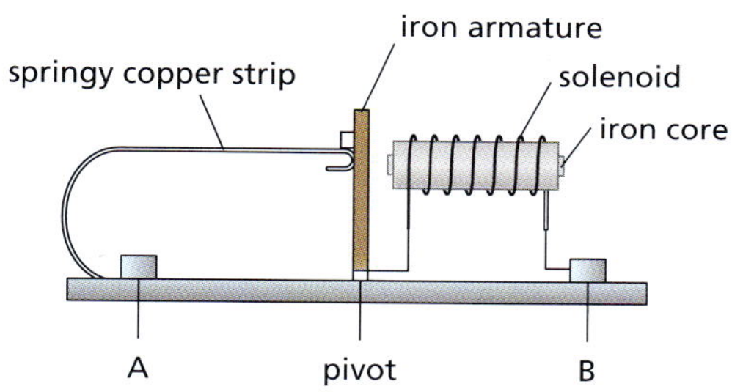

FIG 4.7.3 A simple circuit breaker connected to a circuit at points A and B

In the event of a fault in the circuit, the current passing through the **solenoid** increases. This increases the strength of the magnetic field around the solenoid enough for it to pull the iron armature towards it. When the iron armature moves towards the solenoid, the springy copper strip is released and the circuit is broken.

Interesting fact

Circuit breakers are frequently used in place of fuses to protect domestic mains electricity circuits. They work much faster than fuses and can be reset simply by pushing a switch once the fault has been found and rectified.

Check your understanding

1. Explain how an electromagnet is used in an electric bell.

FIG 4.7.4 An electric bell

Key terms

relay a switch that is operated by an electromagnet

circuit breaker a device that prevents the flow of current in a circuit in the event of a fault

armature a metal part that can move to open or close a circuit

solenoid a cylindrical wire coil with a soft iron core inside

4.8 Induced current

We are learning how to:

- investigate the properties of an induced current

Not only does an electric current create a magnetic field, but when a conductor such as a wire is moved through a magnetic field, an **electromotive force**, or e.m.f. is induced in the wire. If the wire forms part of a circuit, an electric current passes in the circuit.

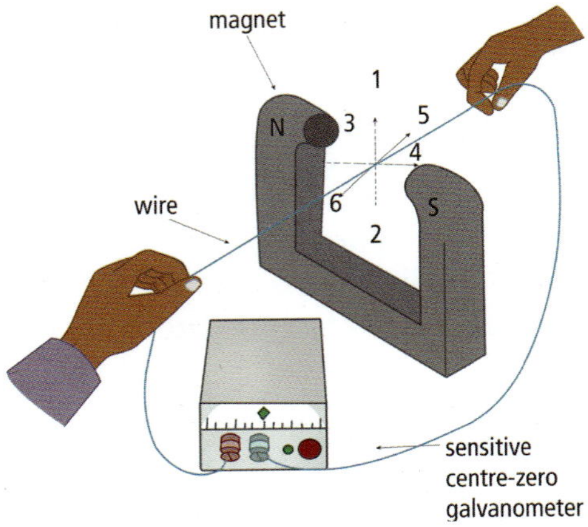

FIG 4.8.1 Inducing a current in a wire

Detecting small currents with a galvanometer

A galvanometer is a very sensitive ammeter that can detect very small currents. The zero point is at the centre of the scale so that currents passing in either direction through a conductor, like a wire, can be detected.

In Fig 4.8.1 a wire has been connected to a galvanometer and held so that a straight section is exactly between the poles of a C-shaped magnet.

Moving the wire in different directions produces different responses on the galvanometer (see Table 4.8.1).

Direction	Response on the galvanometer
1 and 2 (up and down)	The galvanometer deflects in one direction and then in the opposite direction
3 and 4	None
5 and 6	None
Wire stationary	None

TABLE 4.8.1

A current is only induced in a conductor when it cuts across magnetic field lines. If a conductor is held still within a magnetic field, or moves parallel to magnetic field lines, then no current is induced in it.

In the above example, the magnetic field was stationary and the wire was moved within it. It is also possible to induce a current by keeping the wire stationary and moving the magnetic field.

When a magnet is pushed into a coil of wire connected to a galvanometer, the galvanometer deflects in one direction. When the magnet is pulled out the galvanometer deflects in the opposite direction. There is no deflection when the magnet is at rest, even if it is inside the coil, nor when the magnet is moved about inside the coil.

Activity 4.8.1

Inducing a current in a coil

Your teacher will demonstrate this activity.

Here is what you need:

- centre-zero galvanometer
- bar magnet
- coil of wire.

Here is what you should do:

1. Connect the coil to the galvanometer.

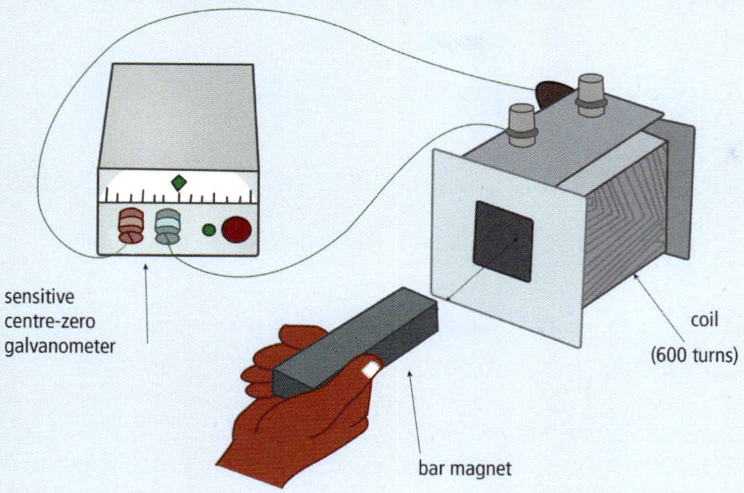

FIG 4.8.2 Inducing a current in a coil

2. Push the magnet into the coil, hold it stationary and then pull it out. Observe what happens to the galvanometer while you do this.
3. Experiment by placing the magnet in the coil and then moving it up and down, and side to side. Observe the galvanometer while you do this.
4. Experiment by varying the speed with which you put the magnet into and out of the coil. Observe the galvanometer while you do this.

Interesting fact

Our knowledge of the relationship between electricity and magnetism owes much to the experiments carried out by the physicist Michael Faraday in 1831.

Key term

electromotive force difference in potential energy that produces a current

Check your understanding

1. Fig 4.8.3 shows a bicycle dynamo. The cylindrical magnet rotates when the driving wheel rubs against the moving cycle wheel. It is used to provide an electric current to power cycle lights.
   - **a)** Explain how the dynamo works.
   - **b)** Explain why a light connected to the dynamo glows more brightly when the cycle moves faster.
   - **c)** Predict and explain what will happen to the size of the induced current when the cycle stops moving.

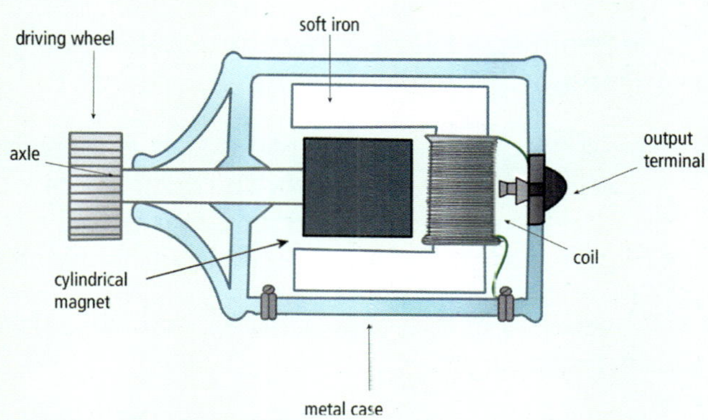

FIG 4.8.3 A bicycle dynamo

Review of Magnetism

- Lodestone, or magnetite, is a mineral composed of iron oxide that attracts objects made of iron. It is a naturally occurring magnet.
- Materials that are attracted to a magnet are described as magnetic, while other materials are non-magnetic.
- Materials that keep their magnetism for a long time are called permanent magnets.
- The law of magnetic poles states: like poles repel, that is, N and N or S and S; and unlike poles attract, that is, N and S or S and N.
- A magnet is surrounded by a pattern of invisible magnetic field lines.
  - Magnetic field lines are conventionally shown to move away from a north pole and towards a south pole.
  - Magnetic field lines never cross over each other.
  - The strength of the magnetic field is shown by the concentration of field lines and is strongest where the field lines are most dense.
  - The magnetic force of a magnet decreases with distance from the poles.
- When a current passes through a wire, a magnetic field is generated around the wire.
- The magnetic field around a wire carrying a current consists of a series of concentric circles that gradually increase in distance between each other.
- Reversing the direction of the current through a wire does not alter the shape of the magnetic field around it, but reverses the direction of the magnetic field lines.
- An electromagnet consists of many turns of wire making a coil or solenoid.
- The first electromagnet was made by the scientist William Sturgeon in 1824. It consisted of about 18 turns of varnished wire wrapped around a piece of iron in the shape of a horseshoe.
- Electromagnetism is sometimes described as temporary magnetism because an electromagnet is only magnetic while a current flows through it. If the current is turned off, the electromagnetic ceases to be magnet.
- The strength of an electromagnet is increased by:
  - wrapping the coils of wire round a core of a magnetic metal like iron or steel
  - increasing the number of turns of wire in the coil
  - increasing the amount of current passing through the coil.
- An electric bell, a relay and a circuit breaker are all common devices that contain an electromagnet.
- An electromagnet can be used to separate iron and steel from other metals in a scrap yard or recycle plant.
- A current is induced in a conductor, such as a wire, when the conductor is moved in such a way as to cut through magnetic field lines.

Review questions on Magnetism

1. Copy Fig 4.R.1 and draw the magnetic field lines around it. Show the direction of these field lines.

FIG 4.R.1 A bar magnet

2. A bar magnet was broken into two pieces.

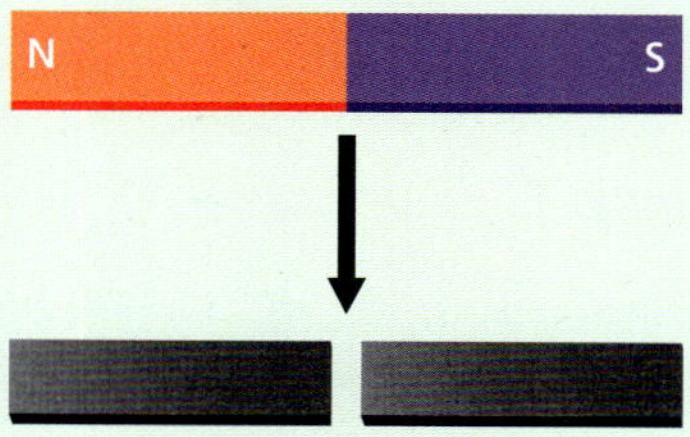

FIG 4.R.2 A bar magnet broken into two pieces

   a) Copy the lower part of the diagram and show the polarity of the new ends formed.

   b) Without using any other apparatus or materials, explain how you could show that both parts of the broken bar magnet have themselves become magnets.

3. Johanna was given four magnets. Her task was to compare how strong they were by counting how many small nails each could lift. Her results are shown in the table.

Magnet	Number of nails
bar magnet	3
C-shaped magnet	6
electromagnet	4
horseshoe magnet	7

TABLE 4.R.1

   a) Which magnet was the strongest?

   b) i) Which magnet could be described as a temporary magnet?

   ii) Explain why.

4. Say whether each of the following will make the electromagnet stronger, weaker or have no effect.

   a) increasing the number of turns of wire

   b) decreasing the current passing through the wire

   c) putting the nail into the coil from the opposite direction

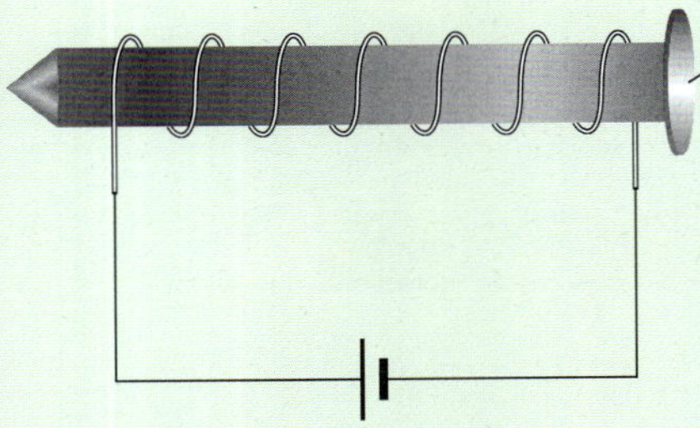

FIG 4.R.3 An electromagnet

5. Here are some statements about magnetic field lines. State whether each statement is true or false.

   a) The closer together they are, the stronger the magnetic field.

   b) They sometimes cross over each other.

   c) They come away from a north pole and towards a south pole.

   d) They are always straight lines.

   e) They are not produced by electromagnets.

6. A student is given three iron bars. The ends of the bars are marked A to F.

FIG 4.R.4 Three iron bars

Two of the bars are known to be magnets and the third one is not. Explain how the student can identify which bar is not a magnet using only the three bars.

7. Fig 4.R.5 shows an electromagnet in a circuit with three bulbs: X, Y and Z.

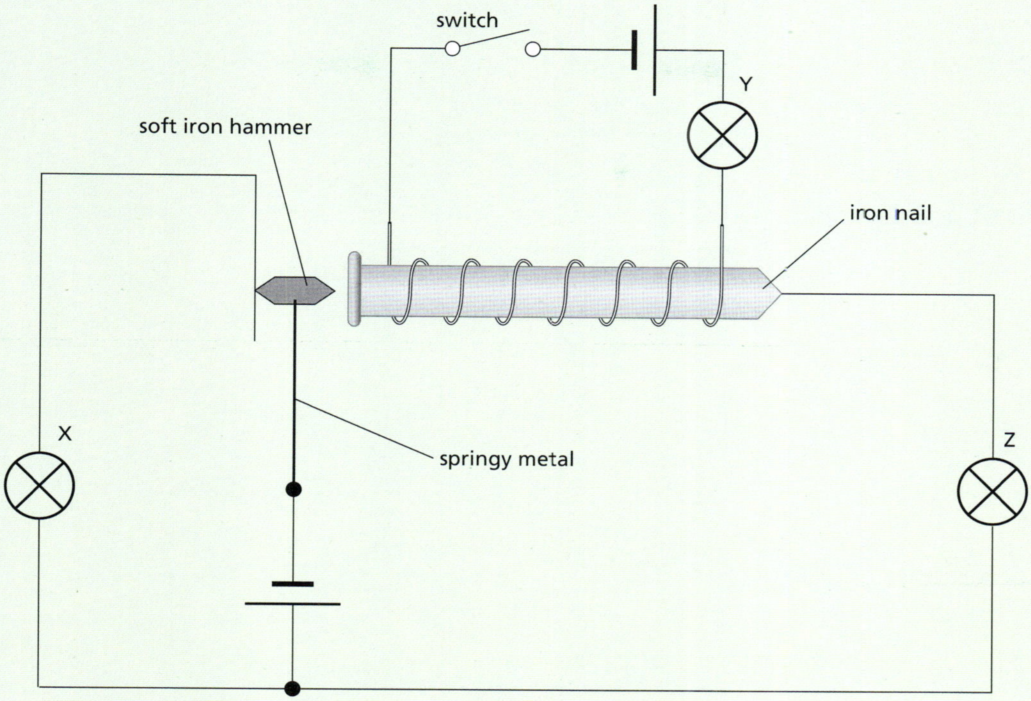

FIG 4.R.5 Electromagnet in a circuit

a) Copy and complete the table by indicating whether each bulb is 'off' or 'on' when the switch is open and when it is closed.

Switch	Bulb X	Bulb Y	Bulb Z
open			
closed			

b) Suppose that the iron nail in the circuit is replaced by a length of wooden dowel of the same diameter. State whether each of the three bulbs will be on or off when the switch is closed. Explain your answer.

Building an ammeter

Progress in understanding science has been made as a result of careful observation and accurate measurement. In order to make accurate measurements, scientists needed instruments like the ammeter.

The ammeters used by early scientists were much different in appearance from those we use today. Early scientists often made their own instruments.

Ammeters measure the current in a circuit. The power of an electromagnet depends on the amount of current passing through it. We can therefore use the magnetic effect of an electric current to measure current.

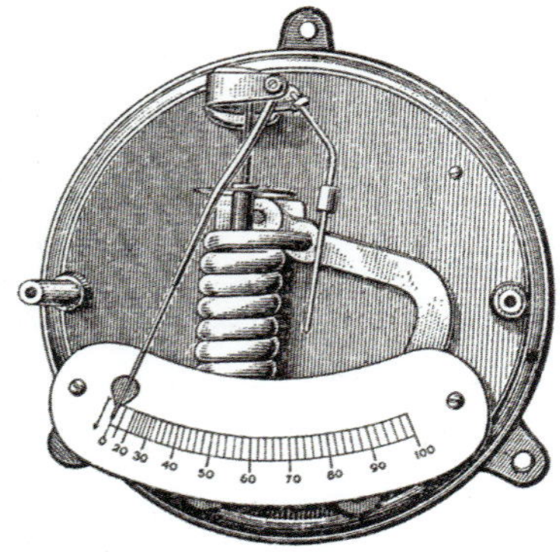

FIG SIP 4.1 Example of an early ammeter

1. You are going to work in a group of three or four to manufacture a simple ammeter using the magnetic effect of an electric current. The tasks are:

- to research how analogue ammeters work
- to design an ammeter that uses the magnetic effect of an electric current
- tconstruct your ammeter
- to calibrate your ammeter
- to check the accuracy of your ammeter against a laboratory ammeter.

a) Research in books and on the Internet how analogue ammeters work. In particular look up:

- moving coil ammeter
- moving iron ammeter.

Don't be concerned about the structure of the ammeters but focus on the principles by which they operate. Increasing the current passing through a coil increases the power of the magnetic field produced by the coil.

b) Apply what you know about electromagnetism and ammeters to designing an ammeter. You will not be able to build anything as sophisticated as a laboratory ammeter. Keep your design simple. Here are a couple of ideas that you might be able to use or modify.

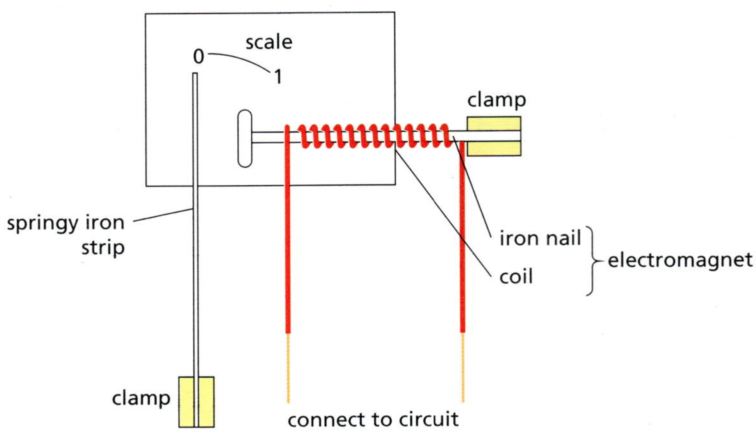

FIG SIP 4.2 Design A for an ammeter

In design A the length of springy iron strip is attracted by the electromagnet and will bend. The greater the current the stronger the electromagnet, and the more the strip will bend. The iron needs to be springy so when the current is turned off the strip will return to zero on the scale.

In design B when a current passes through the coil it becomes an electromagnet and the iron rod is pulled towards it. The greater the current the stronger the magnetic field around the coil, and the greater the pulling force on the iron rod. The iron rod is suspended on a spring so when the current is turned off the pointer will return to zero.

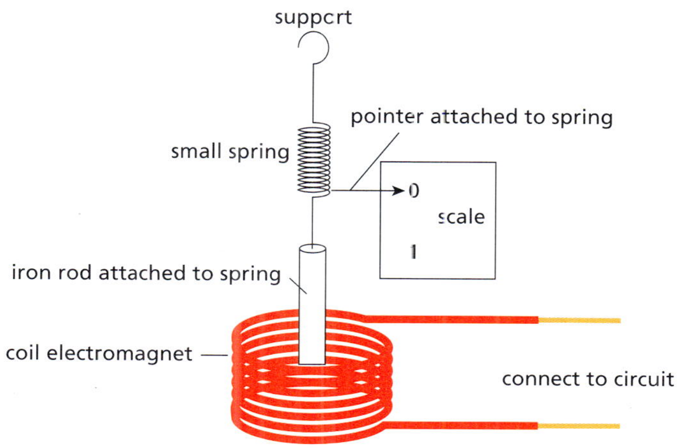

FIG SIP 4.3 Design B for an ammeter

c) When you are satisfied with your design, you should make your ammeter. You might find that during manufacture you need to modify the original design or perhaps incorporate some last-minute ideas. Take some photographs of your ammeter at different stages of manufacture.

d) The scale on your ammeter will need to be calibrated using a laboratory ammeter. Connect your ammeter and a laboratory ammeter in series in a circuit. The circuit should contain a variable resistor so you can alter the current to obtain different values. Use the readings on the laboratory ammeter to draw a scale for your ammeter.

e) Once your ammeter is calibrated, investigate how accurate it is by measuring the current in different circuits and comparing the values you obtain with values obtained using a laboratory ammeter.

f) Prepare an illustrated presentation. You should be prepared to describe how you built your ammeter and discuss the physics involved. You should demonstrate the operation of your ammeter by measuring the current in a circuit. You should also talk about the practicalities of using your ammeter. For example, would it be easy to move from place to place? Can it be easily packed away in a box and brought out when needed?

Unit 5: Thermal energy

5.1 Heat and temperature

We are learning how to:

- distinguish between temperature and heat
- use different scales to measure temperature

Thermal energy is often called heat energy or **heat**. The terms heat and **temperature** are connected but they do not mean the same thing.

Heat is a form of energy. Like all energy, it is measured in joules. Heat energy flows from a hotter body to a colder body.

Temperature is a measure of how hot or how cold something is. When a body gains heat energy, its temperature rises and when it loses heat energy, its temperature falls.

FIG 5.1.1 Temperature is expressed in degrees Celsius (symbol °C) and is measured using a thermometer

Activity 5.1.1

Heat transfer

Here is what you need:

- two beakers
- hot water but not too hot to place your finger in
- iced water.

Here is what you should do:

1. Three-quarter fill a beaker with hot water.
2. Three-quarter fill a beaker with iced water.
3. Place your finger into each beaker in turn and note that the water is hot in one and cold in the other.
4. Leave the beakers to stand on the table throughout the lesson.
5. Near the end of the lesson, place your finger into each beaker in turn.
6. Is the difference between hot and cold as great as it was at the start of the lesson?
7. What has happened to the water in each of the beakers?

When objects are different temperatures, there is a temperature gradient between them. Heat energy is lost by the hotter object and absorbed by the colder object until their temperatures are equal.

Celsius scale

Temperature is measured using a **thermometer**. The units of temperature most commonly used are degrees **Celsius**. This unit has the symbol °C.

The Celsius scale is also sometimes called the **centigrade** scale. The Celsius scale was devised by the Swedish astronomer, Anders Celsius, in the eighteenth century.

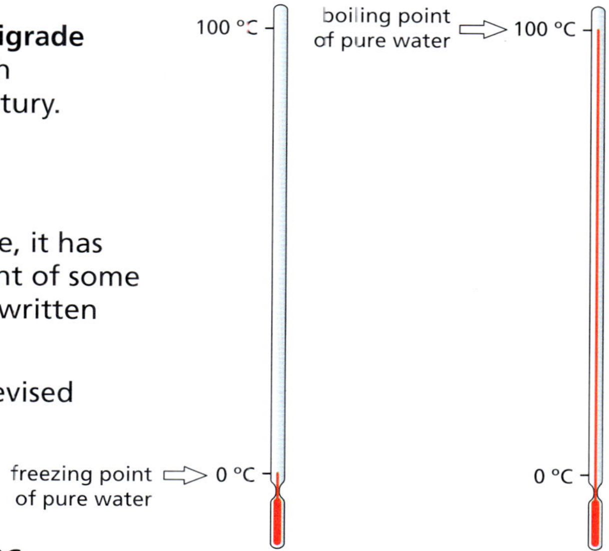

FIG 5.1.2 On the Celsius scale, pure water freezes at 0 °C and it boils at 100 °C, at normal atmospheric pressure (these two temperatures are sometimes called fixed points)

Kelvin scale

Although the Celsius scale is widely used in science, it has one drawback. The melting point and boiling point of some substances are below 0 °C and therefore must be written as negative values.

In order to avoid this problem, a new scale was devised by Lord Kelvin. This scale started at the lowest possible temperature, which is –273.16 °C. This is absolute zero. To make conversion between the two scales easier, the start of the **Kelvin** scale, which is 0 K, is taken to be –273 °C.

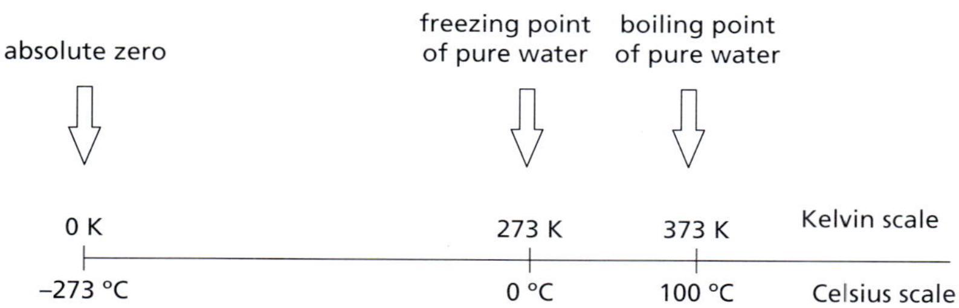

FIG 5.1.3 Kelvin and Celsius scales

Notice that the unit is the Kelvin and not the degree Kelvin, and it is written as K and not °K. The Kelvin is the SI unit of temperature.

Check your understanding

Use the words 'heat' or 'temperature' to complete each of the following sentences.

1 When a body receives .... energy its .... increases.

2 .... is a form of energy while .... is a measure of how hot or cold a body is.

3 When an object is cooled its .... falls because it loses ....

4 The .... of a pond increases during the day because it receives .... from the Sun.

Key terms

thermal energy a form of energy

heat a form of energy

temperature a measure of how hot or how cold something is

thermometer an instrument used to measure temperature

Celsius a unit of temperature

centigrade a unit of temperature

Kelvin a unit of temperature

5.2 Thermometers

We are learning how to:

- distinguish between temperature and heat
- use a thermometer to measure temperature

Measuring temperature

The thermometers most commonly used in the laboratory and elsewhere are called liquid-in-glass thermometers. They consist of a sealed length of capillary tube with a bulb at one end.

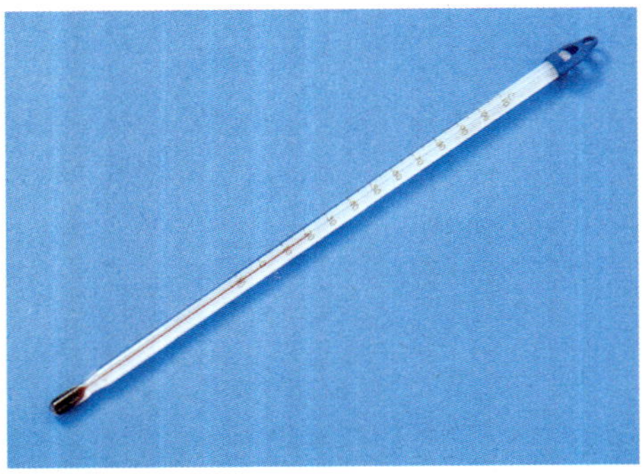

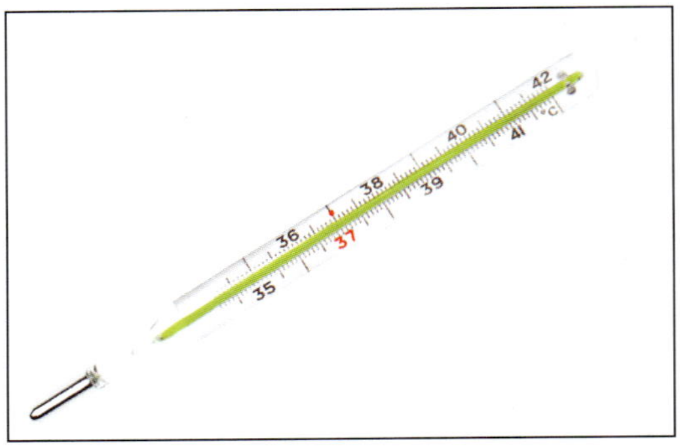

FIG 5.2.1 There are two liquids commonly used in these thermometers: alcohol (ethanol), which is often dyed red, and mercury, which is silver in colour

Care must be taken when reading liquid-in-glass thermometers to obtain accurate readings.

When a liquid is placed in a narrow tube, the surface or **meniscus** is not flat but curved.

The reading on an alcohol thermometer is always taken from the position of the bottom of the meniscus.

The reading on a mercury thermometer is always taken from the position of the top of the meniscus.

Notice that when you read a thermometer, your eye should be level with the top of the liquid.

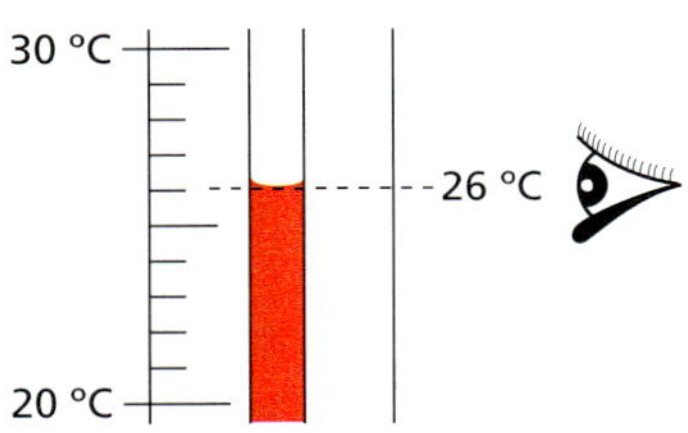

FIG 5.2.2 The meniscus formed by a thread of alcohol is lower at the centre than at the sides

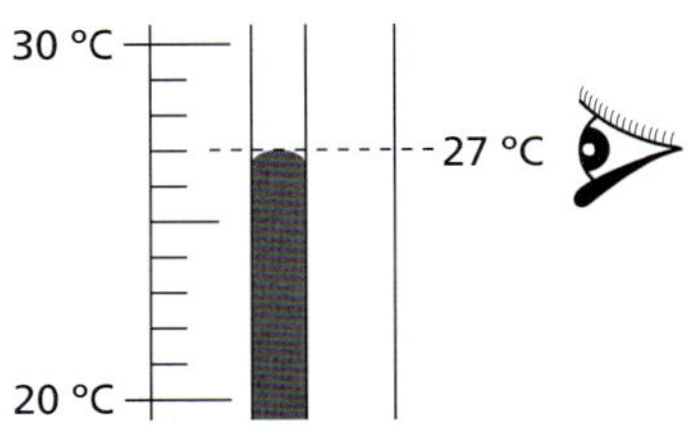

FIG 5.2.3 The meniscus formed by a thread of mercury is higher at the centre than at the sides

Activity 5.2.1

Reading a thermometer accurately

Here is what you need:

- thermometer
- five beakers containing water at different temperatures.

SAFETY

Take care when using hot water. Follow local regulations.

Here is what you should do:

1. Place the thermometer into a beaker containing water.
2. Leave it for a few minutes.
3. Record the temperature on the thermometer.
4. Repeat steps 1 to 3 for beakers containing water at different temperatures.

Check your understanding

1. What is the reading on each of the following thermometers to the nearest half of a degree Celsius?

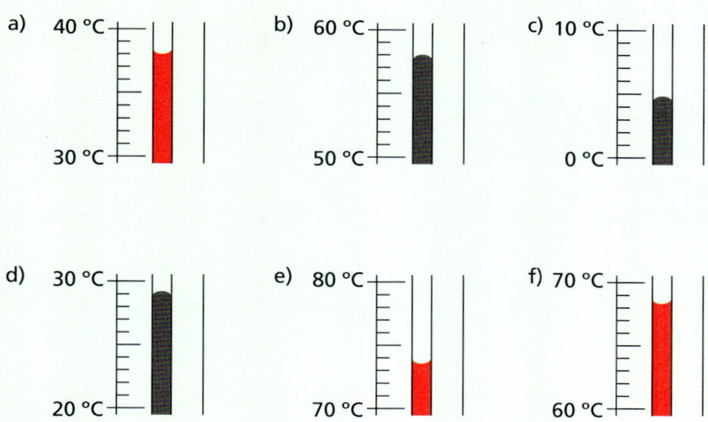

FIG 5.2.4

Interesting fact

A thermocolour thermometer consists of a strip of a special type of substance called a liquid crystal. Some liquid crystals change colour with temperature and are described as thermochromic.

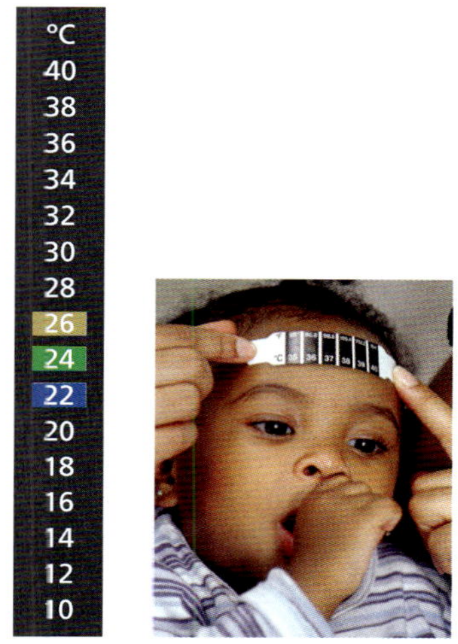

FIG 5.2.5 A simple thermocolour thermometer can be used to show the temperature in a room or in a device such as a refrigerator, or it can be placed on the forehead of a person to measure body temperature

Key term

meniscus surface of a liquid

5.3 Heat transfer – conduction

We are learning how to:

- compare types of heat transfer
- describe conduction as the process by which heat is transferred through solids

Heat transfer

For heat to be transferred, there needs to be a difference in temperature between two points. There are three different methods of **heat transfer**:

- **conduction** – this is the main process by which heat is transferred through solids
- **convection** – this is the main process by which heat is transferred through liquids and gases
- **radiation** – this is the process by which heat can be transferred across open spaces, including the vacuum of outer space.

In real situations, heat is often transferred by a combination of these processes.

FIG 5.3.1 A log fire heats a room by convection and radiation

Conduction

In solids, the particles are held in fixed positions. They cannot move position but they do vibrate.

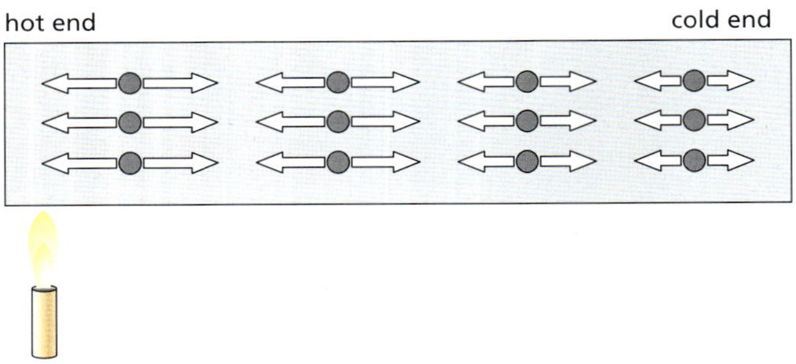

FIG 5.3.2 Conduction in a solid

When a solid is heated, the particles vibrate more energetically. The particles jostle neighbouring particles and in this way movement energy is passed through the solid. This process is called conduction. All solids conduct heat to some extent but some solids are much better **heat conductors** than others.

Activity 5.3.1

Do all solids conduct heat energy at the same rate?

Here is what you need:

- rods of metals and non-metals (such as copper, iron, glass, wood)
- candle wax
- boiling water
- large beaker
- cardboard
- small beaker.

SAFETY

Take care when heating. Follow local regulations.

Here is what you should do:

1. Melt candle wax in a small beaker using an electric heater.
2. Dip one end of each rod in the wax so that it is covered in the wax.

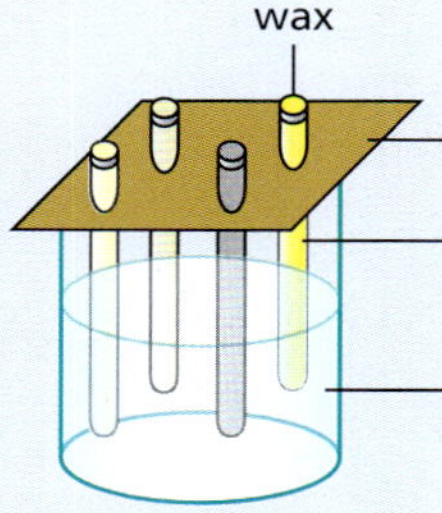

FIG 5.3.3

3. Make holes in a square of cardboard. Make the holes just big enough to push the rods through.
4. Put the rods in the water so the wax-coated ends are out of the water.
5. On which rod did the wax melt first?
6. On which rod did the wax melt last?
7. Which material is the best conductor of heat?
8. Which material is the poorest conductor of heat?

All metals are good conductors of heat while most non-metals are poor conductors. Poor conductors of heat are called **thermal insulators** and include glass, plastics, wood, water and air.

Check your understanding

1. Explain why a toasting fork is made of metal but has a wooden handle.

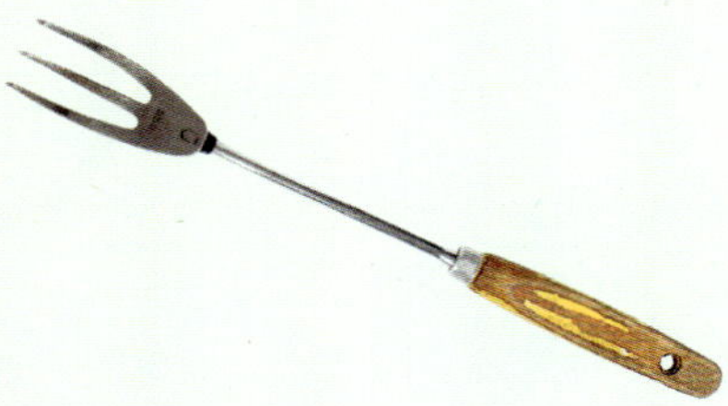

FIG 5.3.5 Toasting fork

Interesting fact

Although diamonds are non-metals, they are the best heat-conducting solids known.

FIG 5.3.4 Diamond conducts heat five times better than copper

Key terms

heat transfer transfer of heat from one place to another

conduction the main process by which heat is transferred through solids

convection the main process by which heat is transferred through liquids and gases

radiation the main process by which heat is transferred through a vacuum such as space

heat conductors objects that transfer heat energy easily

thermal insulators objects that do not transfer heat energy easily

Grade 9

5.4 Metals as conductors of heat

We are learning how to:

- compare types of heat transfer
- explain why metals are good heat conductors in terms of their structure

Why are metals good conductors?

The reason why **metals** are good conductors of heat can be explained by considering their structure.

A metal consists of a framework of tiny particles. Between these particles there are even smaller particles called electrons. These electrons are described as delocalised because they are not held in one place.

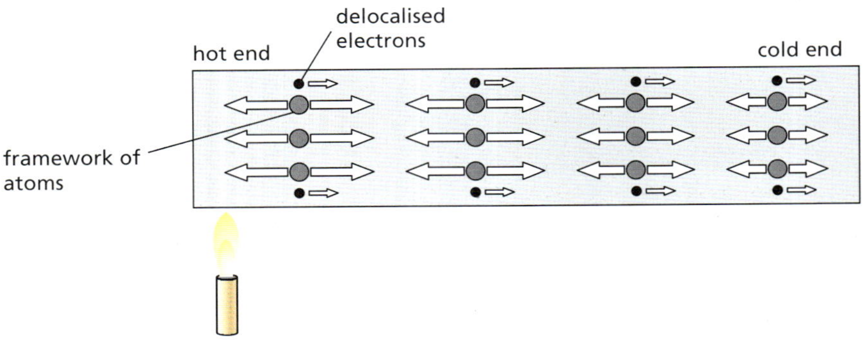

FIG 5.4.1 Metals have mobile electrons

When a metal is heated, the particles vibrate more energetically, just like the particles of any solid but, in addition to this, heating increases the movement energy of the delocalised electrons. The delocalised electrons move quickly and are able to transfer energy through the metal very quickly.

Activity 5.4.1

Do all metals conduct heat energy equally well?

Here is what you need:

- conduction wheel (Fig 5.4.2) or strips of four different metals of equal length
- tripod
- sbeaker
- candle wax
- four paper clips
- heat source.

SAFETY

Take care when heating. Follow local regulations.

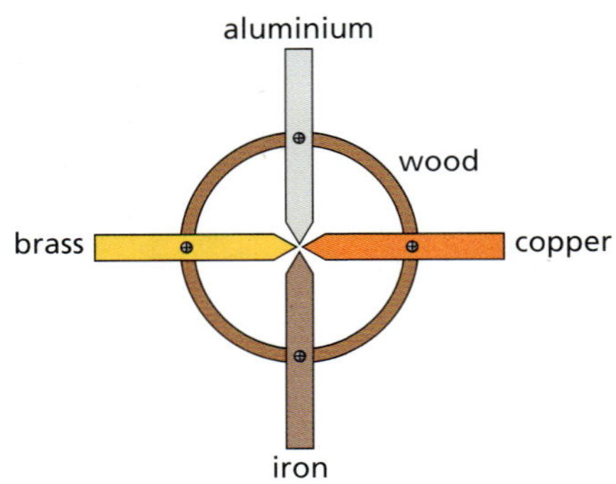

FIG 5.4.2

Here is what you should do:

1. Melt some candle wax in a small beaker.
2. Put a small amount of molten candle wax on the end of the underside of each metal strip in turn, and hold a paper clip to it until the wax hardens.
3. When the metal strip is placed the right way up, the paper clip should hang down from it.
4. Place the wheel on a tripod and heat the wheel at the centre.
5. Record the time taken for the paper clip to fall from each metal strip. Write your answers in a table.

Metal	Time taken for paper clip to fall (s)

TABLE 5.4.1

6. State the order in which the metals conduct heat, starting with the best conductor.

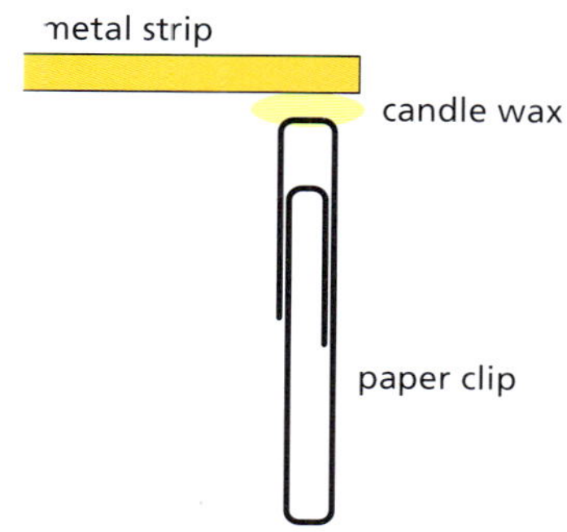

FIG 5.4.3

Check your understanding

1. Some results for the activity described above are given in the table.

Metal	Time taken for paper clip to fall (s)
Aluminium	45
Brass	83
Copper	26
Iron	125

TABLE 5.1.2

a) How can you decide which metal is the best conductor?

b) State the order in which these metals conduct heat, starting with the best conductor.

c) Chromium is a better conductor than iron but not as good as brass. Suggest the time it would have taken for a paper clip attached by wax to fall from a strip of chromium.

Interesting fact

Some cooks prefer using pans that have copper bottoms.

FIG 5.4.4 Copper is a better conductor of heat than aluminium or stainless steel so heat is transferred to the food more quickly

Key term

metal a type of material with delocalised electrons

5.5 Heat transfer – convection

We are learning how to:

- compare types of heat transfer
- describe convection as the main process by which heat is transferred through liquids and gases

Convection

The particles in liquids and gases are too far apart for conduction to take place effectively. Heat is mainly transferred through liquids and gases by another process called convection.

When the particles in a liquid or a gas are heated, they move about more quickly. The particles spread out so they are further apart from each other. The volume of a liquid increases a small amount while the volume of a gas increases by a large amount.

Since the density of a substance is equal to its mass divided by its volume, it follows that when the volume of a liquid or gas increases, its density decreases. This decrease in density is the cause of convection. The less dense substance rises above the more dense substance.

Convection in liquids

It is possible to observe convection taking place in water by introducing a coloured chemical or dye to it.

Activity 5.5.1

Convection in water

Here is what you need:

- beaker
- tripod
- drinking straw
- small crystal of potassium manganate(VII) or dye
- small candle of other heat source.

SAFETY

Take care when heating. Do not touch potassium manganate(VII) with bare hands. Follow local regulations.

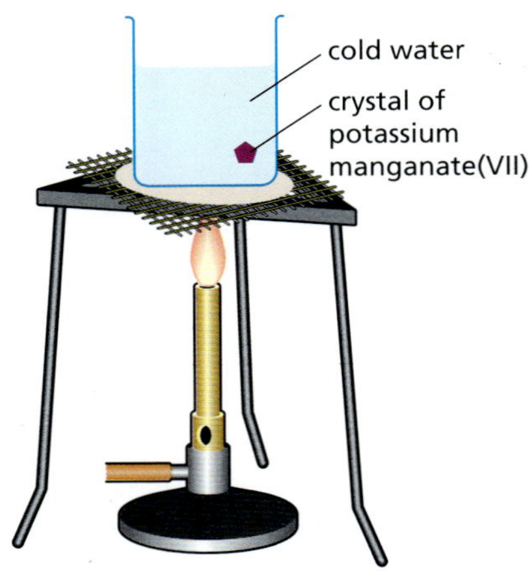

FIG 5.5.1

Here is what you should do:

1. Place cold water into the beaker until it is almost full and place it on a tripod.
2. Stand the drinking straw near the edge of the beaker and carefully drop a small crystal of potassium manganate(VII) down the straw so it sits in the corner of the beaker.
3. Gently warm the water around the crystal of potassium manganate(VII) and observe how the purple colour of this chemical passes through the water.
4. Continue careful heating until all of the water is purple.
5. Draw a diagram to show how the colour passes through the water in the beaker.

The particles responsible for the colour are carried by the water as it moves.

When the water around a coloured crystal at the bottom of the beaker is heated, it becomes less dense and rises, carrying some of the coloured particles with it. Cooler water from the top of the beaker falls down to replace it. Overall, the water rises as it gets warm and falls down as it cools. This movement is called a **convection current**.

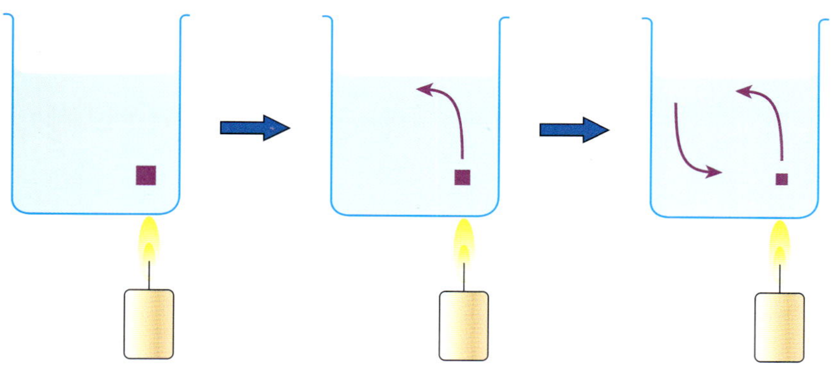

FIG 5.5.2 Convection currents in water

Interesting fact

Convection currents occur on a massive scale in the oceans and in the atmosphere, and are an important factor in determining climate. Differences in temperature at different points in the oceans and the atmosphere result in the formation of convection currents.

In the oceans, convection currents help to circulate nutrients from the ocean floor into shallow water where they promote the growth of organisms.

Check your understanding

1. With the help of a suitable diagram, explain how heat is transferred through water in a heated pan.
2. Describe what convection currents you would expect to notice in the air in a kitchen when there is a hot cooking hob at one side of the kitchen.

Key term

convection current movement of particles of a liquid or gas as a result of convection

5.6 Convection in gases

We are learning how to:

- compare types of heat transfer
- explain convection in gases in terms of changes in density

Hot air

Convection takes place in gases in exactly the same way as in liquids.

Birds of prey know all about convection currents. They spread their wings and rise up on the air that has been warmed by the land.

FIG 5.6.1 Birds of prey can rise to great heights without having to use much energy

Activity 5.6.1

To show that warm air rises

Here is what you need:

- square of thin card
- scissors
- needle
- cotton
- small candle.

SAFETY

Take care when heating. Follow local regulations.

Here is what you should do:

1. Draw a snake on a piece of card by drawing a spiral line out from the centre. Do not make the snake too thin.
2. Cut out your snake so it is in the form of a spiral.
3. Thread a piece of cotton through the head of your snake. Tie a big knot at one end to stop the cotton falling through the hole and tie the other end to the end of a pencil.
4. Hold your snake above a small candle flame.

SAFETY

Do not place the snake too near the flame or it will catch alight and burn.

5. Describe and explain what happens to your snake.

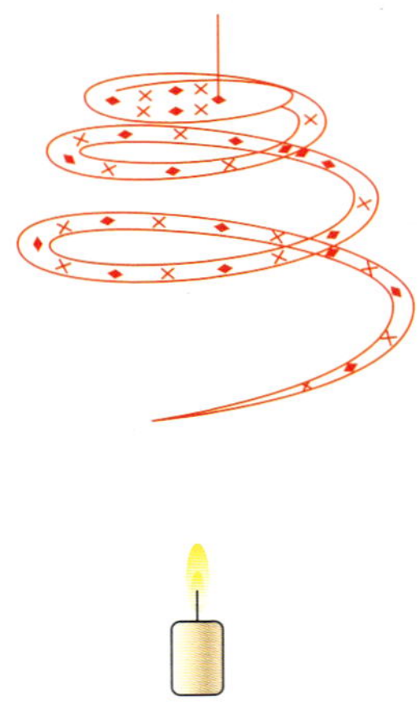

FIG 5.6.2

It is important to appreciate that during convection it is not heat energy that is rising but the heated liquid or gas. The heat energy increases the movement energy of the particles in the liquid or gas. As the movement energy increases, the density decreases and the less dense substance moves above the more dense substance.

In the atmosphere, convection currents are responsible for winds. The Sun does not heat the atmosphere directly. The Sun heats the surface of the Earth and this, in turn, heats the air above it. When the air above the Earth's surface is heated, it expands, its density decreases and it rises, creating an area of low pressure. Cooler, denser air moves in from surrounding areas of higher pressure to replace the rising air.

FIG 5.6.3 You can see convection in a gas by observing a fire: smoke particles from the fire rise because they are carried upwards by the hot air

Check your understanding

1. A smoke box consists of a box that has two chimneys. A candle is burned under one chimney while smoke from smouldering wooden splints is introduced at the other. The smoke box has a glass front so the movement of smoke can be observed.

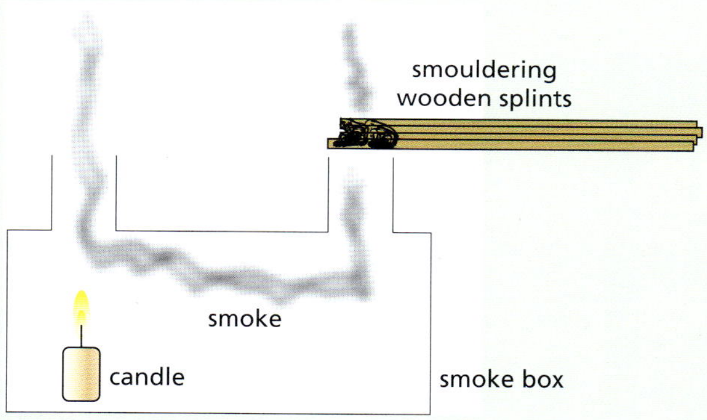

FIG 5.6.4

a) Explain why smoke from the smouldering wooden splints is drawn into the box.

b) Would the same thing happen if the candle was not burning? Explain your answer.

Interesting fact

Convection currents help to keep rooms cool in the summer.

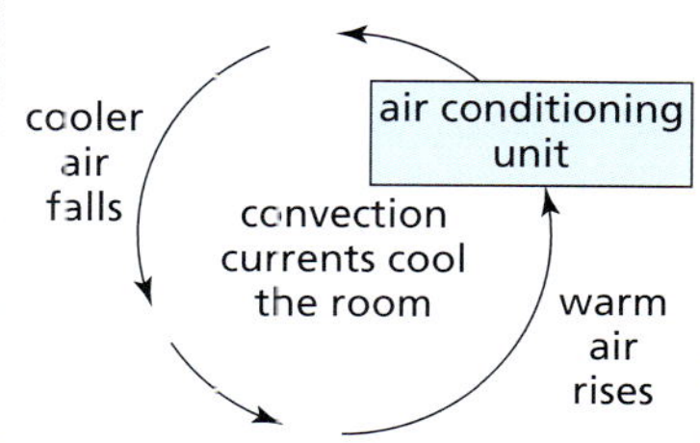

FIG 5.6.5

Warm air rises and passes through the air conditioning unit. The cooler air falls, pushing the warm air upwards.

5.7 Heat transfer – radiation

We are learning how to:

- compare types of heat transfer
- describe radiation as the process by which heat is transferred across a vacuum

Radiation

The Earth and the other planets are separated from the Sun by the vacuum of space. A vacuum contains nothing; no solids, liquids or gases. How does heat therefore travel from the Sun to the Earth and the other planets? There is a third way in which thermal energy can be transferred called radiation.

Heat radiation consists of infrared waves that can travel through a vacuum in much the same way as visible light. Heat radiation is not limited to space. There are many examples of devices on Earth that emit heat radiation.

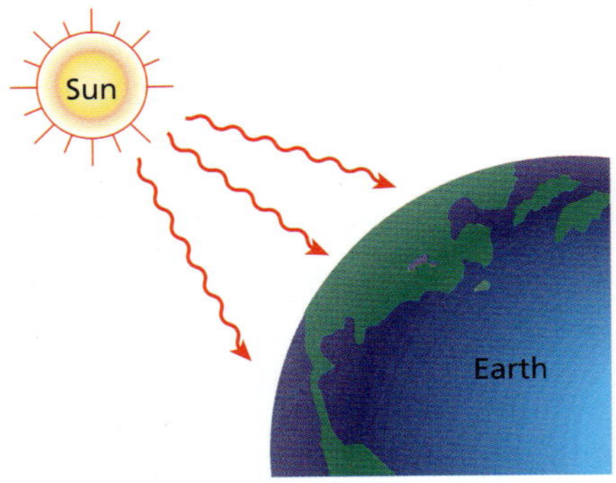

FIG 5.7.1 Heat radiation travels through space to Earth

FIG 5.7.2 An electric fire emits a combination of visible light, which you can see, and heat radiation, which you can feel (the shiny reflector behind the heating element helps to direct the heat radiation into a room)

FIG 5.7.3 Houses in places that get lots of sunshine are often painted in bright colours (the bright colours reflect more of the sunlight than dark colours so the inside of the house is kept cooler)

Tinted glass contains particles of iron that absorb heat radiation. This reduces the amount of heat radiation passing through the window so the inside of the car stays cooler.

FIG 5.7.4 Many cars have tinted windows

Activity 5.7.1

Reducing heat radiation

You will not need any equipment or materials for this activity.

Here is what you should do:

Some different ways of reducing the effect of heat radiation are described in the lesson.

1. Carry out research into some other ways of reducing the effects of heat radiation.
2. Makes some notes and be prepared to share what you have found out in a class discussion.

Check your understanding

1. How do you know that heat does not travel from the Sun to the Earth by conduction or convection?
2. In what form is heat radiation transferred?
3. Why does an electric fire have a shiny metal plate behind the heating element?
4. How can the human body detect heat radiation?

Interesting fact

When it is dark, you cannot see the reflected light from objects but thermal imaging binoculars detect the heat radiation that they emit.

FIG 5.7.5 'Seeing' in the dark

5.8 Radiation and absorption of different surfaces

We are learning how to:

- compare types of heat transfer
- identify good and poor radiators of heat

Emitting and absorbing radiation

Objects can both **emit** and **absorb** heat radiation.

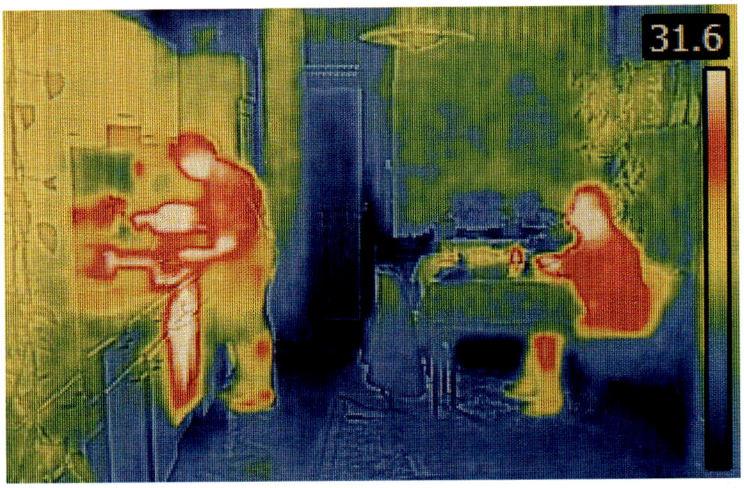

FIG 5.8.1 The heat radiation that is continually emitted by a person makes it possible to 'see' them in the dark using a device that detects heat

FIG 5.8.2 The water in a swimming pool warms up when the Sun shines on it for a few hours

Activity 5.8.1

To investigate if different surfaces absorb heat radiation equally well

Here is what you need:

- two cans of equal size: one shiny and the other painted matt black
- measuring cylinder
- two thermometers.

Here is what you should do:

1. Using a measuring cylinder, fill the two cans with equal amounts of cold water from the same source.

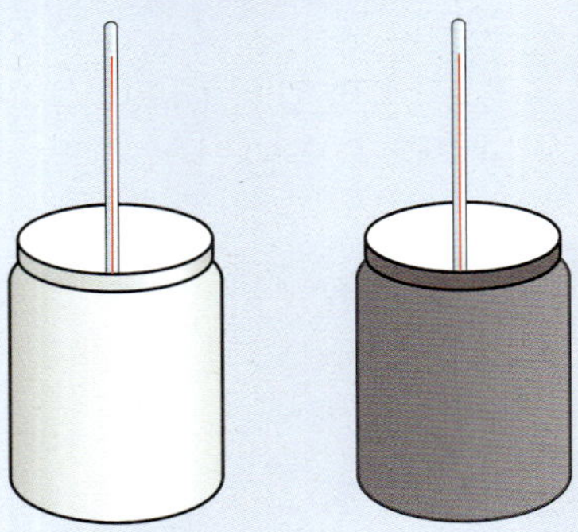

FIG 5.8.3

2. Place a thermometer in each can and measure the temperature of the water it contains.
3. Write the temperatures in a table.
4. Place the cans in the same location somewhere sunny, such as a window ledge.
5. Measure the temperature of the water in each can after 5, 10 and 15 minutes.
6. Write the temperatures in the table each time.
7. Which can was the better reflector of heat radiation?
8. How do you know this?
9. Which can was the better absorber of heat radiation?
10. How do you know this?

Different surfaces do not emit and absorb heat radiation to the same extent.

- Surfaces that are matt (dull and dark) are the best absorbers and the best emitters.
- Surfaces that are light and shiny are the poorest absorbers and the poorest emitters.

Check your understanding

1. Fig 5.8.4 shows details of a solar panel that is used for heating water.

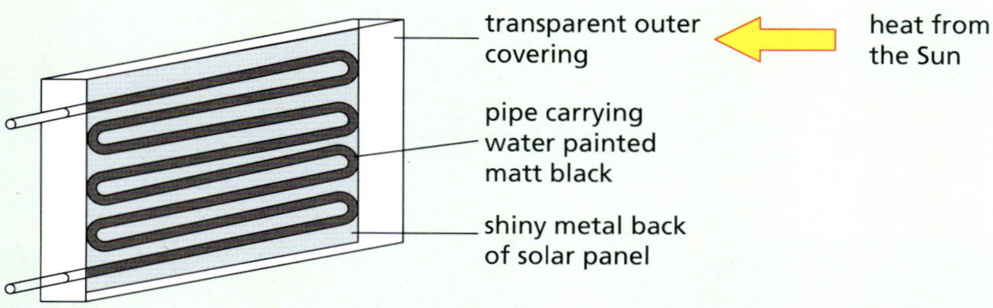

FIG 5.8.4

Explain each of the following features.

a) The panel has a transparent outer covering rather than a wooden outer covering.
b) The pipe is made of copper and not plastic.
c) The pipe is painted matt black.
d) The back of the panel is shiny metal.

Interesting fact

A refrigerator pumps heat from inside to the air outside. The heat is lost through a coil at the back of the refrigerator.

FIG 5.8.5 The coil on the back of a refrigerator is painted matt black so that the maximum amount of heat radiation will be emitted. Heat is also lost from the coil by convection through the air

Key terms

emit give out

absorb take in

5.9 Thermal insulators and conductors

We are learning how to:

- distinguish between thermal insulators and conductors
- appreciate how thermal conductors and insulators are used

Identifying thermal insulators and conductors

A **thermal insulator** is a material that is a poor conductor of heat. Thermal insulators include wood, plastics, ceramics and glass. A **thermal conductor** is a material that conducts heat well. Most metals are good thermal conductors. Cooking pans are often made from metal so that the heat from the hob is conducted quickly to the food.

FIG 5.9.1 Pans often have handles made of wood or plastic so that they can be held while food is being cooked

FIG 5.9.2 Glass is such a poor conductor of heat that a glass blower can hold both ends of a glass tube while heating the middle

FIG 5.9.3 Food stays hot for a long time in a ceramic dish because ceramics are good thermal insulators

Activity 5.9.1

To compare how quickly heat travels through metal and plastic

Here is what you need:

- plastic teaspoon
- metal teaspoon
- cup of hot water
- stopwatch.

SAFETY

Take care when heating. Follow local regulations.

Here is what you should do:

1. Start the stopwatch.
2. Place the teaspoons in the hot water at the same time and hold them there, one in each hand.
3. After two minutes, can you feel a difference in the temperature of the two spoons? Which teaspoon was a better thermal conductor? Which teaspoon was a better thermal insulator?

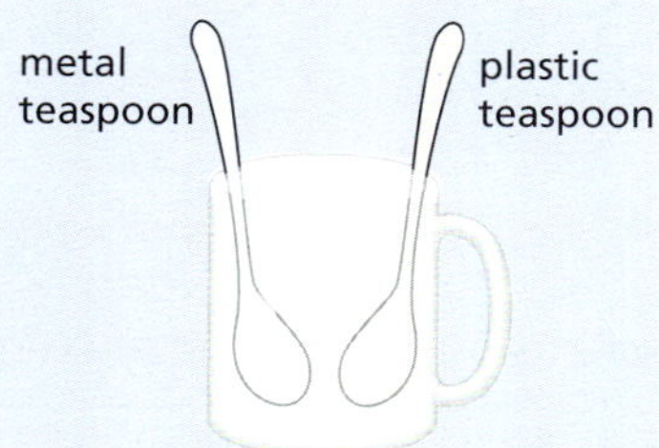

FIG 5.9.4

Check your understanding

1. In an experiment, a strip of white paper was wound round a rod. One half of the rod was brass and the other half was wood.

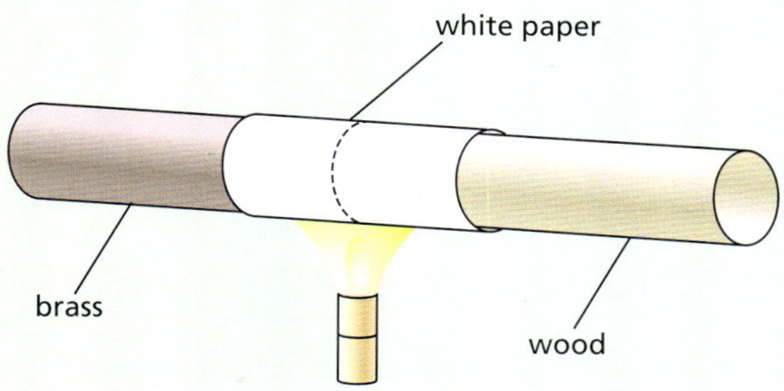

FIG 5.9.5

When the paper was passed over a flame several times, the part wrapped round the wood was scorched but the part wrapped round the brass was not. Explain why.

Interesting fact

Space shuttles used to be used to make regular trips between the Earth and space. When a space shuttle re-entered the Earth's atmosphere from space, the underside got extremely hot due to friction with the air.

FIG 5.9.6

The underside of the space shuttle was covered in ceramic tiles that protected the craft from this potentially damaging heat.

Key terms

thermal insulator a material that is a poor conductor of heat

thermal conductor a material that conducts heat well

5.10 Air as a heat insulator

We are learning how to:

- distinguish between thermal insulators and conductors
- explain the importance of air as a thermal insulator

Trapped air

Heat is conducted through materials as a result of movement energy being passed from one particle to the next. The particles in a gas are much further away from each other than in a solid or a liquid. Gases are therefore poor conductors of heat.

Trapped air can be used as an **insulator**.

Wool keeps animals that live in a cold climate, such as sheep, warm in the winter. The wool can be shorn off the sheep and used to knit woollen jumpers to keep people warm.

Fibreglass consists of many layers of glass fibres. Small pockets of air are trapped between the fibres, giving this material excellent insulating properties.

Bird feathers are sometimes used to stuff duvets because of their insulating properties.

FIG 5.10.1 Wool is a natural fibre that is able to trap air

FIG 5.10.2 Fibreglass is often used to insulate the roof space of buildings to reduce heat loss

Activity 5.11

Which material is the best insulator?

Here is what you need:

- four beakers
- four thermometers
- four types of insulating material
- cardboard
- scissors
- hot water
- measuring cylinder.

SAFETY

Take care when using hot water. Follow local regulations.

Here is what you should do:

1. Cover the bottom and sides of four beakers, each with a different insulating material.

FIG 5.10.3 In cold weather, birds are able to fluff their feathers out to trap air between the feathers and reduce the loss of heat from the body

2. Cut out four cardboard lids for your beakers. Mark the lids A, B, C and D.
3. Make a small hole in the centre of each lid large enough to push a thermometer through.
4. Using a measuring cylinder, put an equal volume of hot water into each of the beakers.
5. Place the lids on the beakers and record the initial temperature of the hot water.

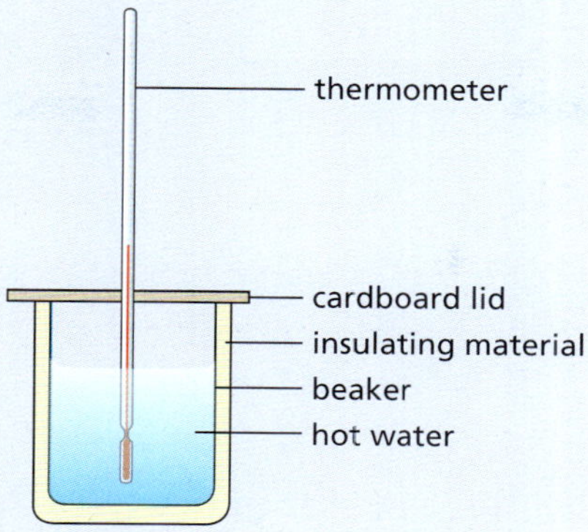

FIG 5.10.4

6. Measure the temperature of the water in each beaker every two minutes.
7. Record your values in a table.
8. Use your data to draw a graph of temperature against time for each insulating material on the same grid.
9. From your graphs, decide which material is the best insulator and which is the poorest.

Check your understanding

1. In some countries, buildings have outer and inner walls. A building can be insulated by filling the cavity with plastic foam containing bubbles of air.

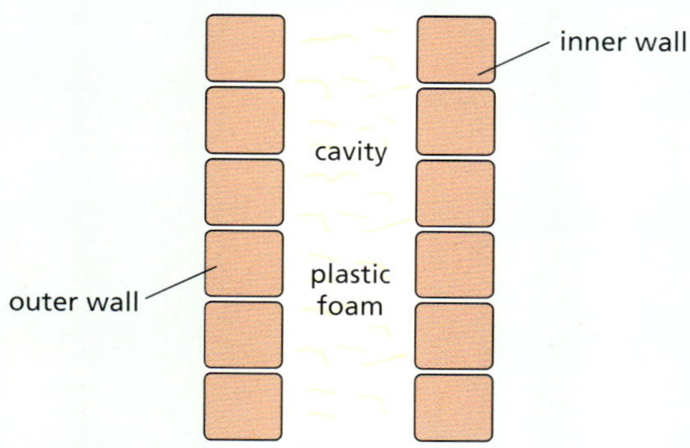

FIG 5.10.5

a) Explain how heat could be lost from the inner wall to the outer wall across the cavity.

b) Explain why filling the cavity with plastic reduces heat loss.

Interesting fact

Hollow synthetic fibres are used in duvets and sleeping bags. The fibres trap air between them but there is also air inside each fibre.

FIG 5.10.6 Hollow fibres are very efficient insulators

Key term

insulator material that reduces heat transfer

5.11 Total heat transfer

We are learning how to:

- distinguish between thermal insulators and conductors
- appreciate the contribution of different processes to total heat transfer

Everyday examples of heating or cooling seldom involve just one method of heat transfer. More often, several processes occur together.

A device that is designed to control the transfer of heat must take into account all three methods of heat transfer.

FIG 5.11.1 In a wood-burning stove, heat is transferred through the metal stove by conduction, the air in the room is heated by convection and the fire also gives out heat radiation

Vacuum flask

A vacuum flask is designed to keep liquids hot on a cold day or cold on a hot day. To do this, it must reduce heat transfer by conduction, convection and radiation.

The vacuum flask contains a glass vessel, which has an inner and outer wall. The air is removed from between the walls to leave a vacuum. The inner surfaces of the vessel are silvered. The flask is sealed by a cork stopper and plastic cap. The table on the next page shows how the features of the vacuum flask reduce heat transfer.

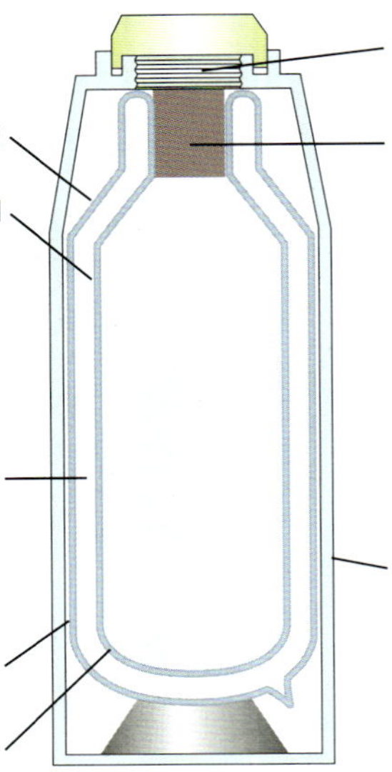

FIG 5.11.2 Inside a vacuum flask

FIG 5.11.3 A vacuum flask keeps liquids hot on a cold day or cold on a hot day

Method of transfer	Feature of the vacuum flask
Conduction	The vessel is made of glass, which is a poor conductor. The stopper is made of cork and plastic, which are also poor conductors.
Convection	The vacuum between the inner and outer walls of the vessel prevents heat loss from the vessel by convection. Convection currents cannot form where there is no air.
Radiation	The inner surfaces of the vessel are silvered. The inner wall reflects heat radiation back into the vessel. This will prevent heat radiation leaving the flask so keeping the contents hot. The outer wall reflects heat radiation from outside the flask. This will prevent heat radiation entering the flask so keeping the contents cold.

TABLE 5.11.1

Activity 5.11.1

Examining a vacuum flask

Here is what you need:

- vacuum flask.

Here is what you should do:

1. Carefully examine the vacuum flask and identify the structures discussed in the lesson.
2. Note the different materials used to make the vacuum flask and consider whether they are thermal conductors or insulators.
3. Look at the mirrored surfaces and consider how reflecting heat radiation reduces the transfer of heat.

Interesting fact

FIG 5.11.4 Cool bags are designed to reduce heat transfer by conduction and radiation

Check your understanding

1. Fig 5.11.5 is a diagram of a refrigerator.
   - **a)** Explain why the refrigerator walls contains insulating material.
   - **b)** Suggest a suitable material for insulating the refrigerator walls.
   - **c)** Name the process by which heat is transferred:
     - **i)** from the fluid through the metal tube
     - **ii)** from the cooling fins into the room.
   - **d)** Explain why the tube is made of metal and not plastic.
   - **e)** Explain why there are fins on the heat exchanger.
   - **f)** Explain why the cooling fins are painted matt black.

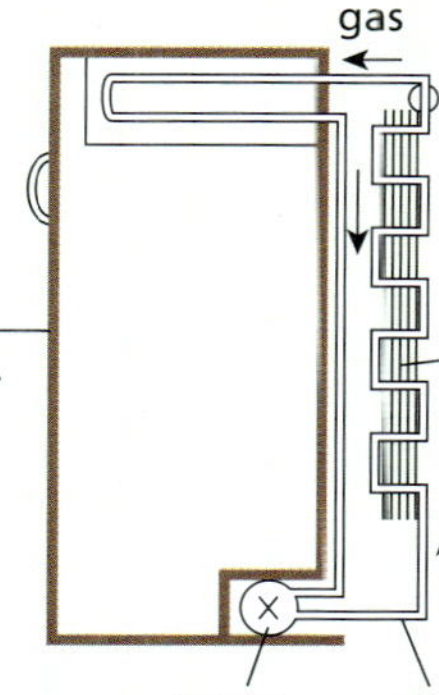

FIG 5.11.5

Grade 9

5.12 Physical properties that vary with temperature

We are learning how to:

- investigate physical properties which vary with temperature

In general all solids, liquids and gases **expand** when heated but not to the same extent.

Expanding solids

Both metallic and non-metallic solids expand when heated. The expansion of metals can be demonstrated using the ball and hoop apparatus in Fig 5.12.1. The metal ball does not fit through the hoop when it is cold, but when the hoop is heated it expands sufficiently for the ball to fall through.

Interesting fact

In the early days of railways, engineers did not think about this and trains were derailed when the lines buckled.

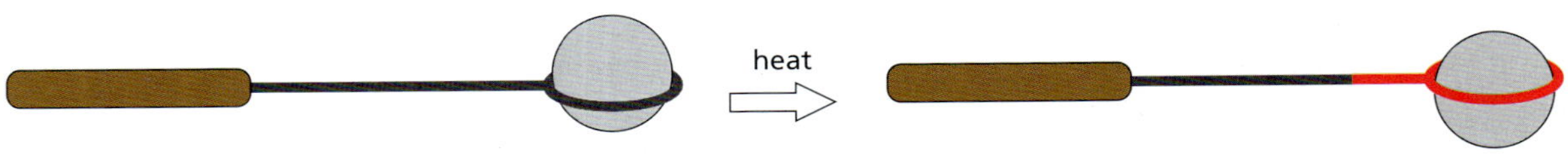

FIG 5.12.1 Metals expand when heated

Large metal structures like bridges expand in hot weather. One end of the bridge is fixed while the other sits on rollers so the bridge can expand without **buckling**. Similarly, railway lines expand in hot weather, so a small gap is left between sections of line to allow for expansion.

FIG 5.12.2 Bimetallic strip

A bimetallic strip is a strip composed of thinner strips of two metals bonded together. One metal expands more than the other when heated causing the strip to bend. Bimetallic strips are used in fire alarms and thermostats, where the bending strip completes an electric circuit.

Expanding gases

Gases expand far more than solids or liquids for the same increase in temperature.

If a flask of air is connected to a U tube containing a coloured liquid, the levels of the liquid in the two arms of the tube change as the air is heated.

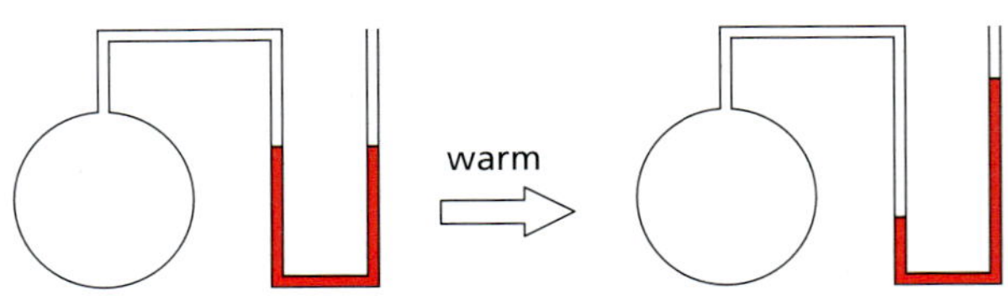

FIG 5.12.3 A gas expands as its temperature rises

Activity 5.12.1

Investigating heat and expansion

Here is what you need:

- balloon
- iced water
- plastic basin
- metre rule
- heat source.

1. Blow up the balloon and seal it by tying a knot.
2. Using a metre rule, measure its diameter.
3. Rub the balloon on your jumper or with your hands to warm the air in the balloon and measure the diameter of the balloon again.
4. Place the balloon in a basin of iced water for several minutes to cool the air in the balloon and measure the diameter of the balloon again.
5. How did the diameter of the balloon change when it was heated and cooled?

Check your understanding

1. Explain what would happen if the air in the flask in Fig 5.12.3 was cooled by placing a flask of iced water around it.
2. Table 5.12.1 shows by how much 1 m lengths of different metals will increase in length per 1 °C rise in temperature.

Metal	Increase in the size of 1 m length of metal per 1 °C rise in temperature / m
Aluminium	0.000 023
Copper	0.000 017
Iron	0.000 012

TABLE 5.12.1

a) By how much will a 1 m length of copper increase if its temperature increases by 5 °C?

b) By how much will a 50m rail line made of iron increase in length if it's temperature increases by 20 °C?

Key terms

expand get bigger

buckling bending and twisting

Review of Thermal energy

- Heat is a form of energy and is measured in joules.
- The temperature of an object is a measure of the amount of heat energy it contains.
- When two objects are different temperature are placed together, heat energy passes from the hotter object to the cooler object until both objects are at the same temperature.
- Temperature is measured using a thermometer. The most common type of thermometer used in laboratories is called a liquid-in-glass thermometer. The liquid is usually alcohol (ethanol) (containing a dye to make it easier to see) or mercury.
- The top of a column of liquid is called the meniscus and it is not flat.
- Temperature is most often measured in degrees Centigrade or Celsius. The symbol for this is °C.
- On the Centigrade scale pure water freezes at 0 °C and boils at 100 °C at normal atmospheric pressure.
- The Kelvin scale is also sometimes used in science.
- There are three different processes by which heat may be transferred: conduction, convection and radiation.
- Conduction is the way in which heat is transferred through solids. Heat increases the kinetic energy of the particles in a solid and this is passed surrounding particles. Metals are good heat conductors, while non-metals are generally poor heat conductors or insulators.
- Convection is the main way in which heat is transferred through liquids and gases. When a liquid or gas is heated its volume increases. As its volume increases its density decreases. It rises away from the heat source, to be replaced by cooler liquid or gas. This results in the formation of convection currents.
- Radiation is the way in which heat is transferred across a vacuum. Heat radiation is sometimes called infra-red radiation.
- Materials that are poor conductors of heat energy are called thermal insulators, or just insulators.
- Air and other gases are poor conductors of heat. Many materials used as insulators, such as fibre glass, rock wool and expanded polystyrene, have pockets of gas trapped inside them.
- In real situations, heat transfer is seldom limited to just conduction, convection or radiation. Heat transfer is more often a combination of two or even all three of these processes.
- Devices used to prevent the passage of heat energy, such as the vacuum flask, must be designed to reduce conduction, convection and radiation.
- Solids, liquids and gases expand when heated.

Review questions on Thermal energy

1. Fig 5.R.1 shows a mercury thermometer.

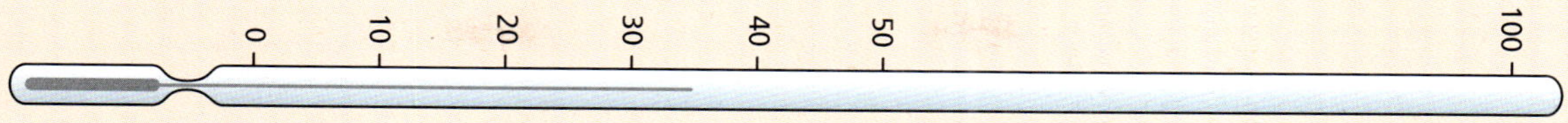

FIG 5.R.1

   a) Copy Fig 5.R.1 and use a ruler to complete the scale between 0 and 100 °C.

   b) Cyclohexane is a chemical that melts at 7 °C and boils at 81 °C. Mark these temperatures on your thermometer using 'M' for melts and 'B' for boils.

   c) In what state is cyclohexane at room temperature?

   d) Apart from being a liquid, state two other properties of mercury that make it suitable for use in a thermometer.

2. Fig 5.R.2 shows a frying pan. It is composed of two materials, X and Y.

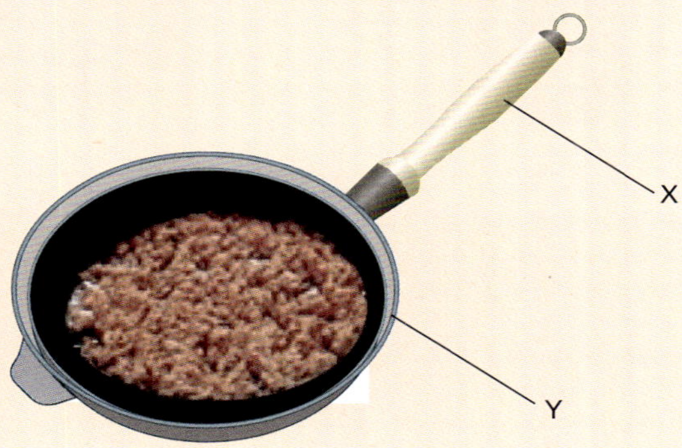

FIG 5.R.2

   a) i) Suggest what material X could be.

      ii) Explain your answer.

   b) i) Suggest what material Y could be.

      ii) Explain your answer.

3. A student filled a thin-walled flask with cold coloured water and connected a capillary tube to it.

She then held the flask in her hands for ten minutes.

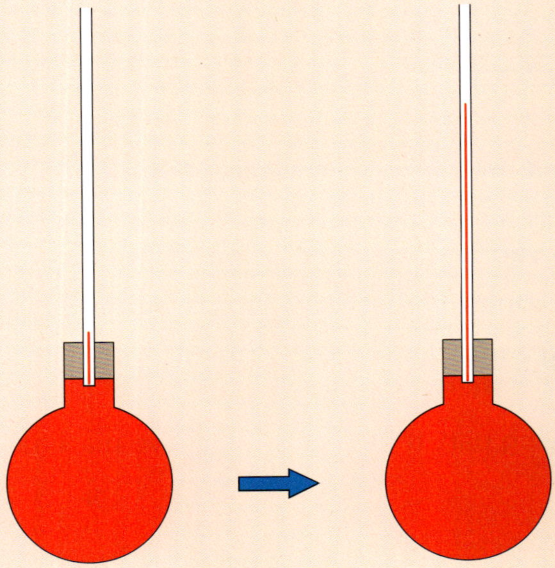

FIG 5.R.3

a) What change was there to the apparatus?

b) Explain this change.

4. Fig 5.R.4 shows ice cubes being heated in test tubes A and B.

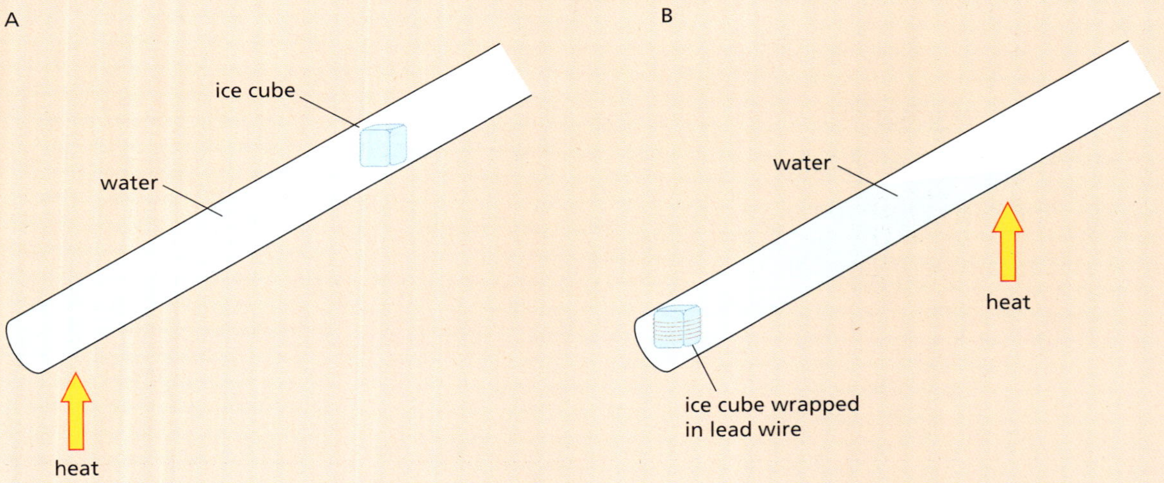

FIG 5.R.4

a) In which test tube will the ice cube melt more quickly?

b) Explain your answer to a).

5. An electric heater was placed exactly between two cans holding equal volumes of water. The cans were identical except that one can was shiny silver (can A) and the other had been painted matt black (can B).

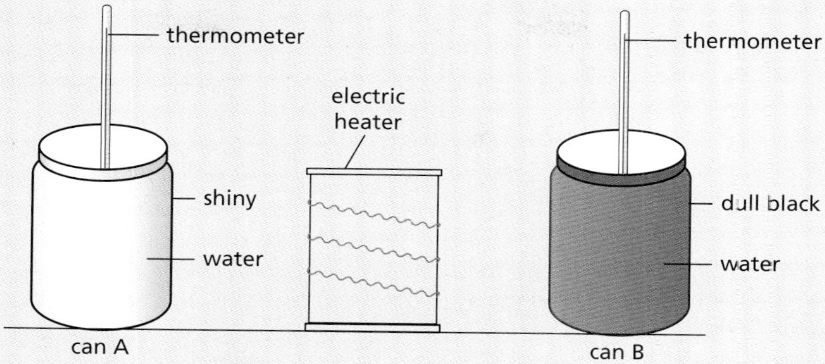

FIG 5.R.5

The temperature of the water in each can was measured every minute for a quarter of an hour. The data collected is given in the table.

Time (minutes)	Temperature of water in can A (°C)	Temperature of water in can B (°C)	Time (minutes)	Temperature of water in can A (°C)	Temperature of water in can B (°C)
0	20.0	20.0	8	24.0	25.5
1	20.5	20.5	9	25.0	26.5
2	20.5	21.0	10	25.5	28.0
3	21.0	21.5	11	26.5	29.0
4	21.5	22.5	12	27.5	30.5
5	22.0	23.0	13	28.5	31.5
6	23.0	24.0	14	29.5	33.0
7	23.5	25.0	15	31.0	34.5

a) State three things that were kept the same to ensure that a fair comparison was made between the cans in this activity.

b) Name the main process by which heat energy is transferred from the electric heater to the two cans.

c) Plot graphs of temperature, on the *y*-axis, against time, on the *x*-axis, for can A and can B on the same grid.

d) Explain any difference in the graphs for the two cans.

Science in practice

Using fuel made from recycled material instead of charcoal

In order to cook food using a coal stove or a jerk pan, a fuel is needed to provide heat.

Traditionally, charcoal is used to cook food on a coal stove. It generally comes in bags of chunks of different sizes.

Although charcoal is made from wood, which is a renewable source of energy, its manufacture damages the environment and creates pollution.

Every year as a society, we produce millions of tonnes of waste paper and cardboard. Much of this can be recycled.

FIG SIP 5.1 Charcoal provides the heat to barbeque food

FIG SIP 5.2 Making charcoal is harmful to the environment

FIG SIP 5.3 Paper and cardboard waste

1. You are going to work in a group of three or four to manufacture briquettes from waste paper and cardboard. The tasks are:

- to research about charcoal currently for sale in your community
- to identify sources of waste paper and/or cardboard that you think might make a good fuel
- to devise a method of manufacturing briquettes from waste paper/cardboard
- to compare how much heat is produced by your briquettes compared to commercially available charcoal
- to compile a report which is to be accompanied by a presentation that includes photographs and/or a video sequence showing how well your briquettes burn.

a) Take a look at the different brands of charcoal for sale in your local market or supermarket. Measure the sizes of the chunks. This will give you some idea what size to make your briquettes. Ask the shop assistant which brand is the most popular. Why does he or she think this is the case? What are the important features of the most popular brand?

b) Identify sources of waste paper and/or cardboard that can be used to make briquettes. Carry out some preliminary experiments to see if some types of paper and cardboard burn more easily than others. For example, is it easier to light a newspaper page or a page from a glossy magazine?

Are there any other issues to be taken into consideration? For example, your briquettes are going to heat food; do some types of paper or cardboard emit noxious fumes that would spoil the food?

c) Make some briquettes.

Some devices are able to compress wet newspapers into 'logs'. When the logs have dried out in the sun, they can be used as fuel.

Can you modify this method to make briquettes?

FIG SIP 5.4 This device turns wet newspapers into logs for burning

d) Devise a way to compare the amount of heat given out by your briquettes with the commercially available charcoal.

It takes 4.184 J of heat energy to raise the temperature of 1 g of water by 1 °C. Perhaps you could heat a given mass of water with a given mass of briquette for a set time and measure the temperature increase. If you repeated this for charcoal but kept the parameters the same, you could make a meaningful comparison of the heat they produce.

e) You should record such things as:

- the composition of your briquettes
- your method of making briquettes
- how your briquettes performed in terms of the amount of heat given out when compared to commercially available charcoal
- whether your fuel stayed alight and glowed like charcoal.

You should take some photographs of video sequences when testing your briquettes.

Use the information you have gathered to prepare an illustrated presentation. You should be prepared to explain why you chose a particular way to make briquettes and to discuss how they performed. You should also discuss possible changes to your method as a result of the experience you have gained thus far.

Science in practice

Making a cool box

Cool bags are used to keep food cool when you go on a picnic.

You could also use a cool bag to keep things warm. The materials used in its construction are designed to prevent heat flow between the inside and the outside of the bag in either direction.

FIG 6.1 A cool bag

The Head Teacher at your school has decided to sell food and drinks to spectators during the school sports day in order to raise funds for new sports equipment. He is concerned that the cost of buying ready-made cool bags will greatly reduce the profits made on selling refreshments so he has turned to his science students for help.

1. You are going to work in a group of three or four to make a cool box out of readily available materials. The tasks are:

- to research about the design of cool bags currently for sale in your community
- to identify readily available materials that you think could be used in the construction of a cool box
- to devise a method of manufacturing the cool box from the available materials
- to compare how efficiently your cool box keeps food and drinks cool compared to a commercially available cool bag
- to modify the design of your cool box to improve its efficiency
- to compile a report which is to be accompanied by a PowerPoint presentation. This should include photographs and/or a video sequence showing the construction of your cool box, and data showing how efficiently it performs.

a) Refresh your memory by looking back through Units 5.3–5.8 on how heat is transferred and through Unit 5.9 on thermal insulators and conductors.

b) Take a look at the different cool bags for sale in your local market or supermarket. What materials have been used in their manufacture? Do the sides feel padded? Can you guess from feeling them what the padding might be? How do these materials relate to what you know about insulators and conductors?

c) What materials are readily available, either for free or at low cost that you could use to make a cool box? Can you obtain cardboard boxes from the local supermarket or perhaps the school office?

What about the materials to fit inside your cool box? For example:

- Can aluminium foil be used to reflect heat radiation?
- Cotton wool and fibreglass as good insulators but may not be readily available. What about waste paper from a shredder? Would that provide sufficient insulation?

Are there any other issues to be taken into consideration? For example, your cool box needs to be light enough to move when it is filled with food. It would be inappropriate to use materials that might taint the food or drink in any way.

d) Make your cool box.

Draw up a plan of what materials you are going to use and how you will use them.

Fig 6.2 shows one possible structure for a cool box. For this design you might consider questions like:

- How thick should the layer of insulation be?
- What insulating material should be used?
- Is there any advantage in having a second layer of aluminium foil attached to the inner layer of cardboard?

Take some photographs at different stages of the construction.

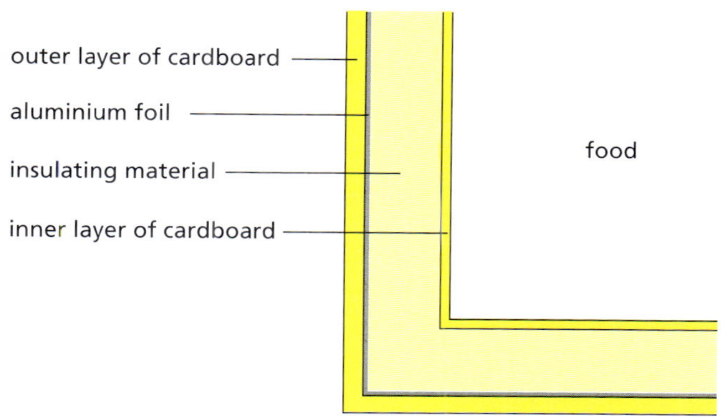

FIG 6.2 Possible design for a cool box

e) Devise a way to compare the efficiency of your cool box with that of a commercially available cool bag. You could place a beaker of iced water into each and monitor how much the temperature rises during one hour.

Use the results of your test to modify your construction and retest your cool box until you are happy with its performance.

Prepare a PowerPoint presentation of the work you have carried out. You should be prepared to explain your design and your choice of materials, and illustrate your method of construction with suitable photographs or a video sequence. You should use the data collected to describe how your cool box performed compared to a commercially available cool bag and say how you modified your design to improve its performance.

Index

A

Absorb, heat radiation, 86–87

Air

as heat insulator, 90–91

hot, 82–83

trapped, 90

Alloy, 51

Amber, 16

Ammeter, 30–31

building, 70–71

Ampere (amp, symbol A), 30, 31

Analogue ammeter, 30

Area, 4–5

of composite shape, 4

estimation, 6, 7

formulae for calculating, 5

of irregular shapes, 6–7

of rectangle, 5

of square, 5

Atoms

defined, 14, 15

structure of, 14–15

B

Bar magnets, 52

Battery, 26, 27

Bimetallic strip, 94

Buckling, 94, 95

C

Car battery, 29

Cells, 26, 27

and lamps, 28–29

Celsius, defined, 73

Celsius scale, 73

Centigrade, 73

Centimetre cubed/cubic centimetre (cm^3), 8, 9

Centimetre squared (cm^2), 4, 5

Ceramic/ferrite magnets, 51

Charcoal, 100

Circuit breaker, 62, 63

Circuit diagram, 32–33

constructing circuits from, 34–35

Components, in electrical circuit, 32, 33

Composite shape

area of, 4

volume of, 9

Conduction, 76–77

Conductors, 22, 23

metals as, 78–79

Convection, 76, 77, 80–81

in gases, 82–83

in liquids, 80–81

Convection current, 81, 83

Conventional current flow, 27

Cool bags/box, 93, 102–103

Copper, 79

Cube, volumes of, 8

Cubic metres (m^3), 9

Current

in circuit, 42–43

induced, 64–65

magnetic effect of, 56–57

measurement of, 30–31

Cylinder, volume of, 8

D

Democritus, 14

Digital ammeter, 30

Displacement

defined, 11

volume by, 10

Double insulated, 25

E

Earth wire, 25

Electrical appliances, rating of, 31

Electrical circuit

building, 26

components in, 32, 33

connecting components in parallel, 38–39

connecting components in series, 36–37

construction from circuit diagrams, 34–35

current, 42–43
direction, 27
measurement of, 30–31
defined, 27
symbols, 32–33
voltage in, 42–43
wires in, 25
Electric bell, 62
Electric current. *See* Current
Electricity, 22, 23
and safety, 24–25
Electromagnet
making, 58–59
strength of, 60–61
uses of, 62–63
Electromagnetism, 59
Electromotive force, 64, 65
Electron flow, 27
Electrons, 23
Electrostatic charge. *See also* Static electricity
defined, 18, 19
problems caused by, 19
uses of, 18
Element, 14, 15
Emit, heat radiation, 86–87
Estimation, area, 6–7
Expand
defined, 95
gases, 94
solids, 94

F

Faraday, Michael, 65
Fibreglass, 90
Fixed value resistors, 40, 41
Friction, 16, 17
Fuel, from recycled material, 100–101

G

Galvanometer, 64
Gases
convection in, 82–83
expanding, 94

H

Heat, 72, 73
Heat conductors, 76, 77
Heat radiation. *See* Radiation
Heat transfer, 72
conduction, 76–77
convection, 80–81
defined, 77
radiation, 84–85
total, 92–93
Hollow synthetic fibres, 91
Horseshoe magnet, 52
Hot air, 82–83

I

Induced current, 64–65
Insulators, 22, 23
air as, 90–91
Irregular shapes, area of, 6–7
Irregular solid
defined, 10
volume of, 10–11

K

Kelvin scale, 73
Kilometres squared (km^2), 5

L

Lamps
brightness of, 36
cells and, 28–29
in parallel, 39
in series, 37
symbols for, 33
Law of magnetic poles, 52–53
Lighting system, 48–49
Lightning, 16, 17
Lightning conductors, 19
Like poles, 53
Liquid-in-glass thermometers, 74
Liquids, convection in, 80–81
Lodestone, 50

M

Magnetic effect of current, 56–57
Magnetic field lines, 54–55
Magnetic field strength, 55, 57
Magnetic material, 51
Magnetite, 50
Magnets
ceramic/ferrite, 51
permanent, 51
poles, 52–53
Meniscus, 74, 75

Metallic structure, 23
Metals
  defined, 79
  as heat conductors, 78–79
Metre squared (m^2), 5
Multimeter, 30

N

Non-magnetic material, 51

O

Ohm, 40, 41

P

Parallel
  connecting components in , 38–39
  lamps in, 39
Particles, 14
  sub-atomic, 14–15
Permanent magnets, 51
  uses of, 62–63
Photocopiers, 18
Plug, examining, 24
Poles, magnetic, 52–53
  like, 53
  unlike, 53
Polythene, manufacturing process, 19
Potential difference, 29, 42, 43

R

Radiation, 76, 77, 84–85
  emitting and absorbing, 86–87
Radioactive isotope, 19
Rectangle, area of, 5
Recycled material, fuel from, 100–101
Refrigerator pumps, 87
Relay switch, 62, 63
Resistance, 40–41
Resistors, 41
  fixed value, 40, 41
  variable, 40, 41

S

Series
  connecting components in, 36–37
  lamps in, 36–37
Socket, examining, 24
Solenoid, 57, 63
Solids, expanding, 94
Space shuttles, 89
Sphere, volume of, 8
Square, area of, 5
Static, defined, 16, 17
Static electricity. *See also* Electrostatic charge
  charging by friction, 16
  charging with, 16–17
  effects of, 17
  investigating, 16–17
Sturgeon, William, 59
Sub-atomic particles, 14–15
Surfaces, absorption of radiation, 86–87
Symbols, electrical circuit, 32–33

T

Temperature, 72, 73
  measurement of, 74
Thermal conductor, 88–89
Thermal energy, 72, 73
Thermal imaging binoculars, 85
Thermal insulators, 77, 88–89
Thermometers, 73, 74–75
  reading, 75
  thermocolour, 75
Total heat transfer, 92–93
Trapped air, 90

U

Unlike poles, 53

V

Vacuum flask, 92–93
Variable resistors, 40, 41
Voltage
  in electrical circuit, 42–43
  measurement of, 42
Voltmeter, 42, 43
Volts (V), 29
Volume
  of composite shape, 9
  of cubes, 8
  of cuboids, 8
  of cylinder, 8
  defined, 8, 9
  by displacement, 10
  of regular shaped objects, 8–9
  of sphere, 8

W

Wires, in electrical circuit, 25
Wool, 90

Acknowledgements

The publishers wish to thank the following for permission to reproduce photographs. Every effort has been made to trace copyright holders and to obtain their permission for the use of copyright materials. The publishers will gladly receive any information enabling them to rectify any error or omission at the first opportunity.

(t = top, c = centre, b = bottom, r = right, l = left)

Cover Lana Po/Shutterstock, p5 Windu/Shutterstock, p10t Yauhenka/Shutterstock, p11 (banana) Nataliia K/Shutterstock, p11 (stone) maxim ibragimov/Shutterstock, p14tl Ratchat/Shutterstock, p14tc Hywit Dimyadi/Shutterstock, p14tr Maurice Savage/Alamy Stock Photo, p16tr Panksvatouny/Shutterstock, p24tl PhotoSerg/Shutterstock, p24tr PhotoSerg/Shutterstock, p25tr John Kasawa/Shutterstock, p25c Krivoshein Igor Alexandrovich/Shutterstock, p25b Sue yassin/Shutterstock, p26br Volodymyr Krasyuk/Shutterstock, p30tr Sciencephotos/Alamy, p30cr Prasolov Alexei/Shutterstock, p40tc Tanasan Sungkaew/Shutterstock, p40rr Jfanchin/Shutterstock, p46tl B Christopher/Alamy Stock Photo, p46br Paul Burns/Getty Images, p47tr Nor Gal/Shutterstock, p48 Phil Degginger/Alamy, p49tr Fedorov Oleksiy/Shutterstock, p49b Peter Sobolev/Shutterstock, p51br Terekhov igor/Shutterstock, p53b Cordelia Molloy/Science Photo Library, p57l Tosnkaanvi/iStockphoto, p57r Science & Society Picture Library/Getty Images, p61 Strelov/Shutterstock, p68tr NEW YORK PUBLIC LIBRARY/SCIENCE PHOTO LIBRARY, p72tl REVOR CLIFFORD PHOTOGRAPHY/SCIENCE PHOTO LIBRARY, p72tr Bart_J/Shutterstock, p73r PAUL WHITEHILL/SCIENCE PHOTO LIBRARY, p74 Erhan Dayi/Shutterstock, p75tr Mevans/iStockphoto, p75b Ffolas/Shutterstock, p77br JoeGough/iStockphoto, p80tr Anirut Krisanakul/Shutterstock, p81tr RicAguiar/iStockphoto, p82c Natalia Barsukova/Shutterstock, p82bl Kovnir Andrii/Shutterstock, p83t Chesh/Alamy, p83br Terry Moore / Stocktrek Images/Alamy, p84tl Dario Sabljak/Alamy, p84tr Romakoma/Shutterstock, p85r Sciencephotos/Alamy, p86tl Warren Price Photography/Shutterstock, p86tr Celiafoto/Shutterstock, p86b Sergiy Zavgorodny/Shutterstock, p87r Jorg Hackemann/Shutterstock, p88tr Pete Pahham/Shutterstock, p88cr DonNichols/iStockphoto, p88br Simon g/Shutterstock, p89br POWER AND SYRED/Science Photo Library, p90tr ARENA Creative/Shutterstock, p90br MidoSemsem/Shutterstock, p91r Ivonne Wierink/Shutterstock, p98tr Littlewitchz223/Shutterstock, p98cr Yann Arthus-Bertrand/Getty Images, p98br ETAJOE/Shutterstock, p99 Clive Streeter/Getty Images, p91r Ivonne Wierink/Shutterstock

Collins

Maximize your CSEC® exam scores

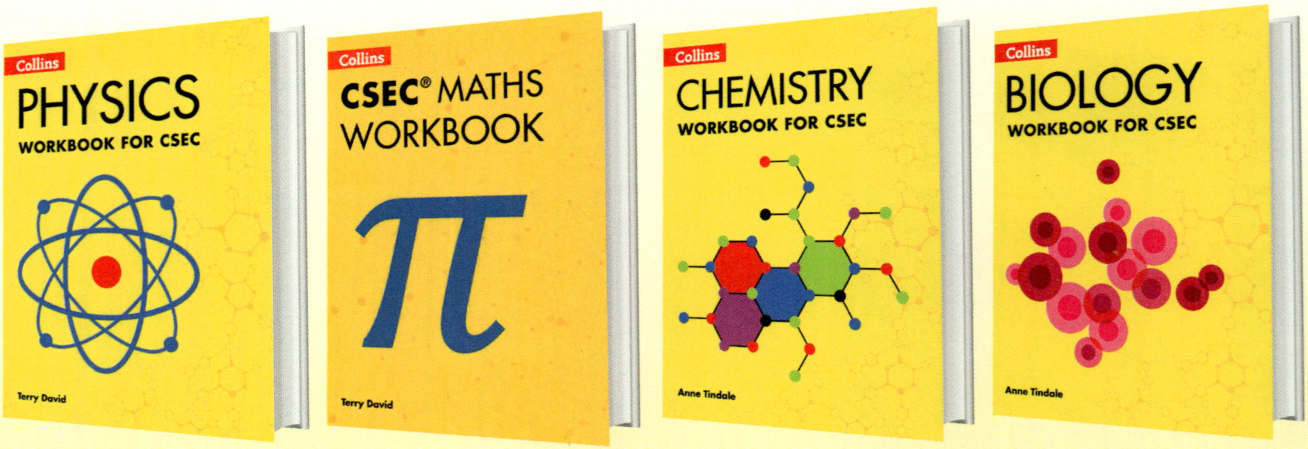

The **Workbooks** provide thorough practice for the structured questions from **Paper 2** of the **CSEC® Examinations** and are a great aid to revision and examination practice.

- Questions and activities on all aspects of the **Biology, Chemistry, Physics** and **Maths** syllabuses
- Clear diagrams, charts and graphs for data analysis activities
- Clear mark allocations to indicate the value of each question part

Collins CSEC® Biology Workbook
978-0-00-811601-9

Collins CSEC® Chemistry Workbook
978-0-00-811602-6

Collins CSEC® Physics Workbook
978-0-00-811603-3

Collins CSEC® Maths Workbook
978-0-00-814739-6

Find out more about publishing for the Caribbean at www.collins.co.uk/caribbeanschools

Collins

Studying the sciences separately at Grade 9

- Developed and written to provide for the **Caribbean** and with full coverage of the **Jamaican** syllabus
- Features special **Science, Technology, Engineering** and **Mathematics (STEM)** activities for each topic
- Provides clear and accessible explanations of each topic
- *'Check your understanding'* sections at the end of each topic to allow teachers and students to assess their progress
- **Workbook** supports the student book by providing further practice of all the key skills

Collins Exploring Physics: Grade 9 for Jamaica
978-0-00-824419-4
Workbook
978-0-00-825208-3

Collins Exploring Chemistry: Grade 9 for Jamaica
978-0-00-824420-0
Workbook
978-0-00-825209-0

Collins Exploring Biology: Grade 9 for Jamaica
978-0-00-824421-7
Workbook
978-0-00-825210-6

Find out more about publishing for the Caribbean at www.collins.co.uk/caribbeanschools